Forewo

Builder; butcher; gallery and gliding club; hatchery; hotel; public house and restaurant. Motoring school, prep school and riding school. Even a rest home!! All these and many more can be found in the Brighton telephone directory prefixed by the name 'Southdown'.

But ask the man in the street — in Sussex or in Portsmouth — what 'Southdown' means to him and I, for one, would be prepared to wager that his answer will be "buses and coaches of course".

For since 1915 the people of East Sussex, West Sussex and the eastern part of Hampshire have been served by Southdown buses when going about their day-to-day business and Southdown coaches have taken them further afield — throughout England and Scotland, Wales and Ireland and to the continent of Europe.

Colin Morris has captured the true flavour of Southdown — the flavour which causes some addicted enthusiasts to declare that the traditional light and dark green paints taste different from cheaper imitations and even refresh the parts that others cannot reach!

This excellent book tells the Southdown story with eloquent words and magnificent photographs and is a fitting tribute to the Company in its 70th anniversary year. It is also a record of how Southdown management and staff have striven to serve their customers. Without devoted staff there would be no service — but without our customers there would be no Southdown.

A. M. Sedgley
General Manager
Southdown Motor Services Ltd
Brighton. January 1985

SOUTHDOWN

Colin Morris

series editor Alan Townsin

The Transport
Publishing Company
128 Pikes Lane
Glossop Derbyshire
Tel: 045.74.61508

for

George Colin Richard Morris

TITLES IN THIS SERIES

Crosville
Aldershot & District
Cumberland
Ribble
Thames Valley
Brighton Hove & District

In course of preparation
Northampton
Dodds/AA Motor Services
Liverpool Buses
Lancashire United/SLT
Manchester Buses
East Yorkshire
Warrington
Preston

COMPANION VOLUMES

British Bus Story 1946-50
British Bus Story 1950's

Best of British Buses
No. 1 Leyland Titans 1927-42
No. 2 The AEC Q family
No. 3 Leyland Tigers 1927-81
No. 6 AEC Regals
No. 7 AEC Regents 1929-42
No. 8 The Utilities
In course of preparation
No. 9 Leyland Titans 1945-84
No. 10 Leyland Lions

Illustrated list of all titles available on receipt of SAE

Typesetting, Artwork and Photography
by TPC Studio, Glossop, Derbyshire

Contents

Introduction

'The Romance of the Southdown; the way it has linked up village to village and town to town, would make a colourful and inspiring book'.

The Bognor Post, 29th September 1934

The above words, written over half a century ago, to commemorate the opening of the coach station in Bognor, put into perspective just how difficult it is to do justice to the history of a company like Southdown Motor Services Ltd — not nineteen years old as it was then, but now celebrating its seventieth anniversary. So much has taken place, so many thousands of working lives have been spent wholly or in-part in its service, and so many outside forces have had their effect upon the manner in which the Company has served the public, that to produce anything less than a full-blown definitive tome seems an inevitable injustice. Perhaps I have just put down a marker for such a book sometime hence.

Meanwhile, the purpose of this volume is to note and record in brief the seventy-year history of passenger-vehicle operation carried out by what is generally accepted in the industry as one of the most distinguished of United Kingdom bus companies and, as such, the text is written in the belief that 'an honest tale speeds best, being plainly told'. Let me say at once that I consider having been asked by the management of the Company to undertake the task a signal honour. Accordingly, I am grateful to the Southdown Enthusiasts Club for having made that recommendation and, in particular, to its founder and number-one card holder, Alan Lambert. Alan, an extremely authoritative author in his own right, has most generously acted as an indispensable local link-man during the research and construction of this volume which was written in far-away Liverpool. He collected and selected material, made numerous tape recordings, compiled most of the appendices and, together with the SEC, has stamped his mark throughout the book.

Although Southdown is seventy years old, its antecedents were beginning to lay the foundations over a century ago. Whether, like California, the climate of the sunny south coast puts everyone in the right mood is difficult to say, but there is considerable evidence that the overall impression of care, courtesy and cheerfulness of its employees conveyed to Southdown's customers began early, and has been nurtured and cherished down the years. That travel by Southdown continues to be viewed as a 'special event', at a time when so many other public 'services' fall woefully short of perfection, speaks volumes for all 'Southdowners'. When many other bus companies are busy dividing themselves into units at the expense of their original identities, it comes as no surprise to learn that, despite speculation about the forthcoming 'magic of the market-place', Southdown's reorganisation of its management and operational structure leaves its respected name intact. What's more, the management is reinforcing that point, as well as celebrating the anniversary, by — among other things — publishing a copy of a 1915 Southdown timetable. How nice; I'm sure Cannon and Mackenzie will approve.

Colin Morris
Liverpool, 1985

Frontispiece The 'Moscow' destination on Leyland Tiger Cub coach No. 1017 of 1955 attracted interest from passers-by as it posed for pictures on Westminster Bridge, with Big Ben as a background. It was one of three vehicles of the type reseated with a capacity of 37 in 1956, though this particular vehicle was the only one allocated to the London-Moscow tour, run by L. W. Morland & Co Ltd of London.

Chapter One: A selection of pioneers

Much has been written in description of the terrain and early modes of communication on the downlands between Portsmouth and Hastings. Roads, scarcely worthy of the name after rain, followed well-worn tracks through the hills towards London or wound their way east-west throughout the length of Sussex. Carriers, coachmen and smugglers alike struggled to shift their burdens as quickly as they were able whilst latter-day highwaymen chanced their luck against them all. In the winter season particularly progress was sometimes bettered by taking to the bed of a shallow stream and, in the middle of the eighteenth century, referring to the depth of the local mud, John Burton was moved to enquire "Why comes it that the oxen, the swine, the women and all other animals are so long-legged in Sussex?"

The need for strong ankles lessened during the next hundred years after local authorities were obliged to attend to the condition of their roads, thus enabling the system of horse-drawn coach routes to be brought to near-perfection; until the coming of the railways creamed off the traffic.

The outstanding personality associated with both forms of transport was William James Chaplin (1787-1859), who horsed and owned coaches and carriages throughout the area, and the best part of England as well. Before becoming chairman of the London & South Western Railway Company, he had taken a look at the possibilities of mechanical traction upon the roads, having travelled as a passenger in Walter Hancock's experimental steam coach **The Infant** which in 1833 set out from London to Brighton. Beaten as much by the state of the road as its own shortcomings, **The Infant** broke down on Handcross Hill, and his horses survived. Seven years later, in June 1840, Frank Hills' 12-seater steam carriage travelled from Deptford to Brighton, and returned to the Elephant & Castle with a full complement of passengers in 3½ hours. At a fare of only five shillings (25p) per person, the trip, although experimental, showed that passengers could be conveyed by steam vehicles at half the expense and at twice the speed of the horse-drawn stage coach.

From that time onward railways increasingly took custom from the coaches until, in the 1890s, many of the roads saw very little traffic. Long distance coach proprietors retreated into the towns, and horses instead provided the tractive effort for hackney carriages, landaus, wagonettes and bus services operating over urban and local routes. In Portsmouth such services were, in the main, provided by a company with interests at various centres in England and Wales; in Worthing and Eastbourne by individual operators; and across Brighton & Hove by similar proprietors who had joined forces to form a syndicate of considerable significance to the Southdown story.

The best-known of the Worthing proprietors, James Town (1826-1912) has left a nostalgic and highly valuable record of the men, horses and stage-coaches of the nineteenth century in his booklet **Reminiscences of the Old Coaching Days**, together with further memories published in the local Worthing press. It would appear that neither coachmen nor 'outside' passengers ever caught cold and, driving some fifty miles per day 'through ten different airs' which produced 'a salutory effect on the system that no medicine can ever do', south coast coach crewmen invariably lived into healthy and great old-age. In addition, these men established a record of getting through on time whatever the weather, of great driving skill and considerable concern for the welfare and safety of their passengers—thus setting a tradition of service and courtesy which remains to this day the hallmark of their present-day successors, Southdown Motor Services Ltd.

The development of the steam carriage upon roadways had been additionally impeded by the constraints upon such traffic imposed by the 'Red Flag' Act which required mechanically powered vehicles to be preceded at a leisurely pace by a man carrying a red flag. By the time the Act had been repealed in 1896 the internal-combustion engine had become a much more promising means of propulsion and of the handful of manufacturers who had produced steam-buses thereafter, only Clarkson's of Chelmsford provided anything of significance to the story of Southdown.

Following the Light Locomotive Act 1896, progress with petrol-engined vehicles was rapid, commencing with a celebratory run from London to Brighton. Wade's Garage of Worthing started with a Cannstatt Daimler and Leon Bollee cars and in 1900 provided the first motor coach in the town, a nine-seater Daimler 'Coventry'. An earlier start with hot-tube ignition vehicles had been made by Arthur Julian at Portsmouth who, in July 1899, introduced experimental motor wagonette services in the town, which led to the inauguration of his Portsmouth & Gosport Motor Omnibus Co Ltd.

WALTER FLEXMAN FRENCH (1856-1925)

Although earlier starts with motorbuses had been made at Edinburgh, in Tunbridge Wells, Bournemouth and Folkestone it is that made between Putney and Piccadilly, London, which is of particular significance here. The proprietor of the MMC wagonettes involved, and who gained his initial experience with motorbuses as a result, was Walter Flexman French, an important contributor to the early history of Southdown Motor Services Ltd—and its first chairman.

Walter Flexman French, the son of a Springfield, Essex, dairy farmer, became an engineering apprentice at the age of twelve. He took honours at the Imperial College of Science & Technology at South Kensington, thereafter spending two years in general engineering and fourteen 'in charge' in locomotive shops. In 1881, he set up on his own account, manufacturing bicycles in a workshop adjacent to his home at Bedford Hill, Balham, and in 1889 became totally self-employed as the maker and vendor 'for cash or on easy terms' of his 'Royal Majestic' cycles, with an output of some two thousand per annum. Frustrated by the red flag requirement, his De Dion car, bought on the proceeds, had but few outings until 1896, when he began to search for ways to aid the process by which the internal combustion engine was to oust the horse, devoting much of his time instead to the motor trade.

To French falls the honour of placing upon the streets of London the first fleet of petrol-engined wagonettes licensed to operate stage carriage services. Started in the summer of 1896 the service between Putney and Piccadilly gave French's son George (later managing director of Maidstone & District Motor Services Ltd) the opportunity of driving the first motorised stage-carriage vehicle over Westminster Bridge. Too frequent stops and starts however ruined the MMC gear-box brasses within a week and difficulty in keeping the back tyres on longer than a day considerably hampered the enterprise. Instead, on 1st April 1901, his seven-strong MMC fleet, some of them still tiller-steered, was switched to a 'country' route - from Clapham Junction and Balham to Streatham, running under the title 'South Western Motor Car Co Ltd'.

When he eventually gave up the production of bicycles and concentrated upon the motor trade he set up anew at 314, Balham High Road, as French's Garage & Motorworks Ltd, and proceeded to become one of the best-known figures in the motor transport world—a 'man of remarkable business acumen, of indomitable perseverance and of the strictest integrity', whose letters and every business move were avidly published in the trade press throughout the first quarter of the 20th century.

This then was the man who for two years from 1904 was to move house and home to Worthing to take up the managership of the recently-formed Sussex Motor Road Car Co Ltd—one of the direct progenitors of Southdown, whose first chairman he was later destined to become.

ALFRED DOUGLAS MACKENZIE (1870-1944)

Although French was its first chairman, Southdown Motor Services Ltd was for many years synonymous with what turned

Two of the most influential figures in the story of the origins and character-moulding of Southdown can be seen in this picture dating from 1914. Douglas Mackenzie, in characteristic nautical-style peaked cap, stands near the driver and Alfred Cannon is the tall figure standing among passengers on the first motor coach tour from Sussex to the Lake District in 1914, the picture being taken at the Keswick Hotel. 'Sussex Tourist Coaches' was the fleetname used by Worthing Motor Services Ltd for such purposes and the green Daimler CC-type has bodywork of Mackenzie's 'slipper' design—its Isle of Wight registration, DL 705, having been taken out by his office on the island. WMS was perhaps the most direct of the predecessors which merged to form Southdown Motor Services Ltd in 1915. Percy Lephard, co-founder with Mackenzie and Cannon of Wilts & Dorset Motor Services Ltd, stands by the nearside front wheel.

W. F. French

A. D. Mackenzie

A. E. Cannon

out to be the lifelong partnership of 'Mac and Freddie' - of Alfred Douglas Mackenzie and Alfred Edward Cannon, respectively traffic manager and general manager of Southdown and joint founders of one of its three constituent companies, Worthing Motor Services Ltd.

Whilst photographs of Alfred Cannon are few and far between, Douglas Mackenzie, in his earlier days both moustached and trim-bearded with naval-style peaked cap planted squarely upon his head, is depicted in numerous photographs of vehicles taken in the first twenty years of the century, in most instances well-placed to listen intently and knowledgeably to the engines which he nursed and loved throughout his life.

In many respects, Douglas Mackenzie was an enthusiast—methodical, cautious and reluctant to throw anything away. Of Ross & Cromarty stock, but born in Kensington, he had travelled to school by train when, according to his own account, he collected locomotive numbers and, long before 'spotters' had it all laid out for them in booklets, devised a system of classification and sequencing to enlighten the task—the first example of that orderliness which was to make Southdown one of the best recorded, ticketed and numbered of bus companies.

The youthful Mackenzie was apprenticed to a firm of marine engineers in Sunderland and undertook several voyages as a trainee ship's engineer. Trained upon steam engines, in the autumn of 1898, he was appointed to the secretarial staff of the National Traction Engine Owner's Association, which group required on their staff an engineer to advise its members about the suitability and strength of the country roads and bridges in their area. Mackenzie travelled the country from North Lonsdale in the Lake District to Looe in Cornwall, looking in particular at bridges. So narrow were most of the country lanes that he was obliged to use a bicycle to reach what he quickly realised was some of the most beautiful countryside in England. Apart from rescuing from almost certain destruction a granite clapper bridge in the Looe Valley, his unique two-year tour was to have a lasting effect. First, in his subsequent Worthing days, Mackenzie and his bicycle became something of a legend but, more importantly, his first-hand contact with such terrain, unknown at that time to the majority of Englishmen, caused him to look "forward to the time when I could get hold of a motor bus chassis sufficiently reliable to send on tours of 500 to 1,000 miles and thus enable others to enjoy the scenery and country I had come to love and value".

To this experience there may be traced the impetus which led to the provision of an extensive programme of long-distance tours for which Southdown was to become particularly famous.

Meanwhile, in 1901, Douglas Mackenzie joined Allen's of Cowley, Oxford, where he managed a fleet of 80 steam-powered vehicles, including lorries, threshing and ploughing engines and steam-rollers. Never in good odour with the local authorities, because of their weight, his engines made eight-inch-deep ruts in the weak surfaces of the roads as far afield as Salisbury Plain and Southend. Mackenzie realised that the lighter, more easily-managed internal combustion engine had the greater future.

Despite a journey across the Vale of the White Horse, between Wantage and Oxford on one of the earliest of motor cars—and that made astride the bonnet holding his overcoat down on either side to stop an easterly gale blowing out the flame in the car's porcelain ignition tube—the future director and traffic manager of Southdown felt it necessary to up-date his knowledge of petrol engine development by attending trials conducted by the Auto-Cycle Club and the Motor Yacht Club at Southampton. On one such trial on Southampton Water, he encountered a Thornycroft motor cruiser driven by producer gas—a system in which no gas was made unless by the suction of the engine which was actually going to use it. This too was to provide experience which would help him cope with the gas propulsion of buses imposed by wartime restrictions upon petrol.

His introduction to motorbuses was made in 1904, when he was employed by the Motor Traction Co Ltd which had started running double-deck petrol buses in London as early as October 1899. By 1905, he had set up as a consulting engineer, first in The Strand and then at 109 Victoria Street, Westminster, becoming much sought-after in the emergent omnibus industry. The Glasgow & South Western Railway Co, Clacton-on-Sea Motor Omnibus Co and the Isle of Wight Express Motor Syndicate were among the firms which engaged him to advise upon vehicle policy and maintenance. For the Clacton firm he designed what he always claimed was the first motor charabanc. Mounted upon a Leyland X chassis this was the first of his famous slipper-type vehicles later to become familiar along the south coast. Additionally, its name, **Swiftsure**, was applied to the vehicle in a beautiful cursive script which owed much to Mackenzie's own neat handwriting and was the prototype for the elegant fleetname **Southdown** applied to the Company's coaches.

The consultant now began to move into the actual management of bus companies and, in March 1907, he was appointed to conduct the affairs of the Isle of Wight Express Motor Syndicate Ltd which he undertook from his office at Westminster. At the time this firm was running a fleet of six Milnes-Daimler charabancs—and was already ailing. Fortunately, the company had actually operated three buses in Portsmouth and Southsea the year before Mackenzie arrived and had done some excursion work to Goodwood Races under the watchful eye of a travel agent called Frank Bartlett.

All then was not lost when, in December 1907, Mackenzie was also called upon to act as receiver when the Isle of Wight Express Motor Syndicate was wound up, for Bartlett was to become the key man in the actual establishment of Southdown at Portsmouth. In addition, Mackenzie kept open an office at Ryde, Isle of Wight, whence he was able to register (and re-register) vehicles for use in his latest, and what turned out to be his permanent venture, in Sussex. For, in April 1907, Alfred Douglas Mackenzie agreed to become full-time general manager of the Sussex Motor Road Car Co Ltd— a position vacated the previous year by Walter Flexman French. To act as engineer, he wooed away from the London Power Omnibus Co Ltd his assistant during his earlier consultancy days, the tall, amiable and 24 year-old Alfred Edward Cannon.

ALFRED EDWARD CANNON (1883-1952)

A native of Sandford-on-Thames near Oxford, Alfred Cannon (1883-1952) had served his apprenticeship as an engineer in the Great Western Railway works at Wolverhampton—an engineering pedigree as good as any other. Yet it was Cannon's flair for management, coupled with Mackenzie's intimate knowledge of what machines could be expected to do, that was to establish the correct balance of innovation and caution which later established Southdown Motor Services Ltd as one of Britain's foremost bus companies.

Mackenzie, incidentally, did not much like the word **bus. Motorbus** he positively hated, considering it to have "a nasty Yankee flavour that is distinctly unpleasant to British palates...". Instead, he called all his buses **cars.** And such was the character of the man, over 40 years after his departure most of the Southdown staff call them that also. The 'Telex' name for Worthing Motor Services Ltd, 'Mobus', later inherited by Southdown, can only be a joke played upon him by Alfred Cannon.

FRANK BARTLETT (1860-1933)

Frank Bartlett's retirement dinner was held at the City Hall, Portsmouth, in February 1933. It was attended by the Lord Mayor of Portsmouth and all the aldermen of that city. Douglas Mackenzie and Alfred Cannon attended for Southdown Motor Services Ltd and every employee of the company subscribed to his farewell present. Bartlett was neither a director of the company nor a man with engineering experience. He was in fact the first manager of Southdown's Portsmouth depot, in which district he had spent most of his life. Yet his contribution to the presence of Southdown Motor Services Ltd in eastern Hampshire was considerable, having acted as the anchor-man in Portsmouth during Mackenzie and Cannon's attempts to link Brighton with Portsmouth.

Born at Ryde, on the Isle of Wight, Bartlett became a telegraph messenger in 1872, was a junior clerk until 1877 when he joined the Ryde Pier Company as a booking clerk at the Isle of Wight Joint Railway Companies' offices on the Esplanade and Pier Head stations. He transferred to Portsmouth in 1884, opening the booking offices at Fratton and the one-time Southsea railway station. After a period back at Ryde and six years spent in Canada, he took up local excursion, steamship and emigration work in his own office at 67 Commercial Road from March 1903.

By now a well-established travel agent, Bartlett contacted Clement Rutlin, the secretary of the Isle of Wight Express Motor Syndicate Ltd in 1905 and convinced him that, following the Corporation's decision to concentrate upon its tramways, there was a gap in the market which could be filled by motorbus stage-carriage services and excursions. Although this project ceased and the Isle of Wight company foundered, he thus met Mackenzie and handled the affairs of the Sussex Motor Road Car Co Ltd, and later Southdown, for him until the establishment of a proper depot in the town, when he became the local manager. Comparatively unsung though Bartlett may have been, Southdown's Portsmouth depot owes its origin to his foresight.

Mackenzie in typical pose, looking thoughtfully towards the engine as the driver appears to be engaging him in conversation—maybe the classic complaint, "pulling poor, guv'nor"—as the conductor looks on. CD 397 was an ex-Brighton, Hove & Preston United 1905 Milnes-Daimler transferred to Southdown, but the radiator is not of the original type, being a BH&P replacement.

The first motor bus in Brighton was this Milnes-Daimler 24hp with the early style of low-mounted radiator, which arrived in December 1903 and was registered CD 103, entering service with the Brighton, Hove & Preston United Omnibus Co Ltd. It did not survive long enough to be included in either of the transfers of ex-BH&PU rolling stock, to the newly-formed Southdown in 1915 or to Thomas Tilling Ltd when BH&PU sold out the following year, but Milnes-Daimler buses played a key role in the development of motor bus operation in the area. The Daimler in the title was the German concern founded by motor vehicle pioneer Gottleib Daimler and Milnes was the tramcar builder, responsible for the bodywork on early examples.

Chapter Two: The constituent companies

1: THE 'COUNTRY' SERVICES OF THE BRIGHTON, HOVE & PRESTON UNITED OMNIBUS CO LTD.

Upon its formation in 1915, three companies contributed toward what was to become Southdown Motor Services Ltd. Worthing Motor Services Ltd and the London & South Haulage Co Ltd were completely absorbed, whilst the Brighton, Hove & Preston United Omnibus Co Ltd subscribed its country routes, excursion and tours together with the personnel, rolling stock and premises concerned, retaining as an operating entity its stage carriage services in Brighton and Hove which, in turn, were sold to Thomas Tilling Ltd the following year. Founded in 1884, the BH&PUOC was the oldest of the three companies involved.

The London Brighton & South Coast Railway ran its first holiday-makers' excursion into Brighton Terminus in 1884; an event which attracted hotel omnibuses to the station thereafter. Within ten years some 300 horse-drawn conveyances were being licensed by the Hackney Carriage Sub Committee. From among their number there developed recognisable omnibus services upon the London pattern and by 1875, their main assembly point was Castle Square, whose residents complained to the Watch Committee of the obstruction to their businesses and the touting of the conductors. The council promised to "take steps to remedy the evil".

In 1877 the several bus proprietors were joined by William Mayner of Westbourne Villas, Aldrington, whose omnibus business in Birmingham had been purchased the previous year by the newly-formed Birmingham Tramways & Omnibus Co Ltd. Mayner ran buses to Cliftonville and between Church Road, Hove and Kemp Town. He and William Taylor Beard, who operated from Church Road to Castle Square and also from North Street to the terminus station, jointly employed a timekeeper to start the buses. Sworn in by a special constable, the unfortunate man's duties extended from 8.30am to 10.30pm each day for 23 shillings (£1.15) per week, the majority of which was paid by Beard.

In addition, there were Walter Tilley's 'Brighton Busses' which operated three services from Castle Square, to Lewes Road, Stamford Avenue and to Tilley's own 'Race Hill Inn', from 1867-1901; the buses owned by A. E. Elliott running between Kemp Town and Cliftonville and Henry Thomas' to Tivoli Gardens. Thomas is of particular note for he also ran 'out of town' from the Terminus station and Castle Square along the breezy cliff top eastward to Rottingdean in competition with Welfare's bus.

As early as 1878 road traffic in Brighton and in Hove had reached such proportions that the shopkeepers in Western Road claimed that it was already inconveniently crowded and 159 of them signed a petition to council protesting against the further issue of licences to omnibus proprietors—a seemingly counter-productive move, for there can be little doubt that tradesmen along the line of route benefited from the growing traffic. The council continued to grant the 5 shilling (25p) licences fairly generously, however, but warned Walter Tilley that after the licensing year 1881-82 they would refuse to license omnibuses of the great size now run by him between New Road and the Cemeteries. Applications by newcomers were nevertheless turned politely aside.

One such attempt—to run small one-horse buses on numerous routes in Brighton—was made by the Brighton General Omnibus Co Ltd in the summer of 1884. Yet it got no farther than the slimmest of toe-holds, before its efforts were pre-empted by a corporate enterprise which was to gather up into a powerful syndicate most of the larger independent operators.

On 11th October 1884, the Watch Committee of Brighton Corporation learned that the Brighton, Hove & Preston United Omnibus Co Ltd had been formed, which proposed to purchase all the omnibuses presently owned by William Taylor Beard, William Mayner, Henry Thomas and Walter Tilley. The committee quickly informed the new company that it would be pleased to license the omnibuses anew in the name of the company. Indeed, it would have been surprisisng if it had not, for Alderman J. L. Brigden and John J. Clark were members of the board of directors, which also included Major-General W. R. E. Alexander, E. A. Eager, S. Ridley and G. Tatham. Beard became traffic and general manager of the new undertaking, Frank Smith its secretary and, in the early days, Eager was its accountant. William Mayner moved to Seafield Mews, Hove, and set himself up anew as a jobmaster. In the event, Tilley seems to have withdrawn from the scheme and continued to run his own buses along Lewes Road until he was put out of business by the Corporation's tramcars, whereafter he concentrated upon his motion-picture-house.

The grand collection of vehicles and equipment included some 30 omnibuses and 150 horses which were considered to be "considerably above the average of Omnibus horses" and housed initially in the constituent proprietors' premises—including Upper St. James's Street and Conway Street, Hove. W. T. Beard and Frank Smith (later the first secretary of Southdown Motor Services Ltd) were installed in the offices at 6 Pavilion Buildings, "the most advantageous spot for collecting the fares from conductors", and set about increasing routes and reducing headways throughout the two neighbouring boroughs. During 1886-87 Brighton alone licensed 42 of its buses, despite fears that the vehicles would "disturb daily services in many places of worship" and spoil the local cab drivers' livelihood, the high price of corn and hay,

This 1906 Milnes-Daimler, CD 495, was one of six which passed to Southdown from the Brighton, Hove & Preston United 'country' fleet in 1915, although all of the vehicles concerned had by then received single deck 'coach' bodywork. It is seen here in earlier days as a double-decker with Birch 34-seat bodywork.

a recent epidemic of influenza "among the visitors and cab-horses" and complaints that buses would bring "excursionists into the best part of town; an annoyance from which residents have happily not yet suffered".

In its horse-bus days the company established the practice of painting either the destination or abbreviated route in large letters on the side panelling beneath, or latterly instead of, the company name in full. It was not until 1891 at least that all its omnibuses were fitted with proper staircases, instead of makeshift ladders, thus permitting lady passengers to travel 'on top'. The 'decency panels' which then required placing across the upper rails—lest a shapely ankle distracted other drivers—provided additional revenue in the form of advertisements for national and local businesses.

Urged on by its director E. A. Eager, the BH&PUOC began seriously to consider the possibility of motor traction in 1901. In the event, it was not until December 1903 that it purchased a motorbus, just as registration number-plates were being issued for the first time. CD 103 was a 24hp Milnes Daimler double-decker, the first of 25 purchased between then and 1909. In 1906 the BH&PUOC began to build some of its own motorbus bodies at its Conway Street premises—an example later adopted by Worthing Motor Services Ltd. E. A. Eager was now the company's 'superintendent and engineer', pressing onward with motorbuses despite the public disquiet felt when, on 12th July 1906, a London-based **Vanguard** double-decker en-route for Brighton came to grief with the loss of ten lives on that same Handcross Hill which had stopped Hancock's steam-coach 73 years before. It was Eager who jumped into the driving seat of a BH&PUOC Milnes-Daimler, and collecting doctors, stretchers and appliances, drove to the scene of the accident and took four surviving casualties to hospital in his bus. Eastbourne Corporation, the first municipality in the world to run motor omnibuses under parliamentary powers (from April 1903) had cold feet about them in 1906 and asked Eager to come along and advise the Motor Omnibus Committee of that town on "matters connected with the management of motor omnibuses".

Subsequent correspondence makes it perfectly clear that Eastbourne tentatively invited the BH&PUOC to take over the Corporation's bus operations, but the directors came to the conclusion that they should "keep it to the area of Brighton for the present".

In Brighton itself, if the clip-clop of horses hooves and the jangling of the brasses had disturbed the peace in 'the best part of town', that was nothing compared with the racket set up by a Milnes-Daimler in full stride. The combination of creaking wooden wheels and rack and pinion final-drive produced a continuous chatter not unlike an amplified magpie; it echoed off shop fronts with great effect and did even better off the paling fences which abounded in the more select areas. Taking the renewed and more vigorous bout of complaints extremely seriously in January, 1909 the BH&PUOC invested in a Hallford Petrol-Electric bus and three Electrobuses, each capable of travelling some thirty miles without recharging. These placated all but the most hardened complainants and the company purchased all the remaining vehicles of the London Electrobus Company in 1910, placing eight of these in service together with another new Electrobus and three more Hallford-Stevens petrol-electrics, thus providing a fleet of sixteen extremely quiet vehicles.

Eager kept a list of comparative mileages covered by petrol and electric-powered buses from 10th January 1909 to 28th October 1916, until the town services were purchased by Thomas Tilling Ltd. At their zenith in 1911, battery electric buses had covered 277,164 miles against 492,140 by the petrol engined vehicles. In 1915 the figure was 230,406 against 714,807. The petrol-engined bus had been seriously challenged but not defeated.

Earlier, the Company had made an attempt to lease or otherwise secure powers to run tramcars over the tramways laid down by Brighton Corporation from 1901. The most that BH&PUOC brought away from that discussion was permission to affix some of its 'stop-tablets' to the anchor poles. Clauses in the Brighton Corporation Act 1903, however, did safeguard the routes of the Company should the Corporation themselves decide to run motorbuses. This of course they did not do in the lifetime of the BH&PUOC or indeed until many years thereafter. Indeed, the Corporation stuck to its tramcars and entered into a series of co-ordination agreements with the Company.

Competition with a different tramway, however, the Brighton & Shoreham Tramways Co Ltd, was another matter. Started in July 1884 with steam-hauled trams, which quickly failed, it relied upon horses stabled in its depot at Halfway House, Southwick, to haul the cars between Shoreham and Westbourne Villas, Hove. The rest of the journey into Brighton had been provided by the horse-buses of William Taylor Beard in the last few months of his independent days. Despite their first manager's earlier agreement with the tramway and the fact that it had been purchased by the British Electric Traction Co Ltd with a view to electrification, BH&PUOC now considered it fair game and proceeded to draw off most of the passengers on this route with their far speedier buses, and finally killed it off in 1912.

The Brighton, Hove and Preston United Omnibus Co Ltd began placing single-decked coach bodies on its Milnes-Daimler additions to the fleet and running them along the coast route to Worthing in December 1905 in competition with the Sussex Motor Road Co Ltd. Additionally, between then and World War I, it applied for licences to run to Rottingdean, to Hurstpierpoint and to Lewes; curiously, however, there seems to be neither documentary nor testimonial evidence that these were taken up. In 1911, it decided to purchase a pair of Straker Squire coaches and commenced running a series of day excursions from Brighton, building up a fleet of fourteen coaches-cum-single-deck buses and two charabancs. The following January, it entered into an agreement with Worthing Motor Services Ltd—Mackenzie and Cannon's successors to SMRC—whereby the

Straker Squire was another make originally based on a German design, in this case the Bussing, though by the time this example on the U-type chassis was built for BH&PU in 1911, the Bristol-based concern claimed to have largely adopted British components. The body, built by Christopher Dodson, had a strong resemblance to Worthing Motor Services' contemporary practice, with stepped seating arrangement derived from, though less extreme than, the original Mackenzie 'slipper' designs. It was lettered for the Brighton-Worthing service, abandoned by agreement with WMS a few months later. The vehicle, identifiable by the chassis number 810 carried locomotive-style on a plate on the bonnet side, was registered CD 1441, and passed to Southdown.

Worthing company retired from the Worthing-Brighton road completely in return for five per cent of BH&PUOC's revenue taken on the route and for concessions to the east of Brighton (see this chapter: part 2).

Requisitions made by the War Office at the commencement of World War I reduced this 'country' fleet to eight vehicles. Like operators in other seaside resorts at the time, the company turned again to horses for its excursion work and acquired some taxis as well. In July 1914, it bought as a going concern, staff included, the business of Brighton jobmaster F. J. Mantell (later to become traffic manager of the Southdown company's Brighton area), together with his licences to ply for hire from the Front, and run to the Devil's Dyke—a service that BH&PUOC had intended to run with motor buses five months previously. By the time the Brighton, Hove and Preston United Omnibus Co Ltd was ready to make its contribution to the newly-formed pool in 1915, it also owned four horse charabancs, two victorias, sixteen landaus, five six-horse charabancs, each with its own name, one four-horse coach and one light covered-van, together with nineteen horses the army didn't want—and three motor taxi-cabs.

In order to take over the eight coaches, fourteen spare bus bodies and two country routes of the BH&PUOC, the new company had to swallow hard and buy this equine collection lock, stock and barrel—and Douglas Mackenzie "didn't know one end of a horse from another".

2: WORTHING MOTOR SERVICES LTD

If not the oldest of the companies involved in the merger, at least Worthing Motor Services Ltd contributed most in terms of rolling stock, routes, personnel and, above all, its dynamic duo—Mackenzie and Cannon. It is also the best documented and recorded. In many ways it could be said that the founding of WMS' predecessor, the Sussex Motor Road Car Ltd in 1904, marks the real genesis of Southdown Motor Services Ltd.

The last days of the stage coach, the reign of the horse bus, and the emergence of motor traction in the Worthing district has been admirably chronicled by Henfry Smail (1948) in his **Coaching Times and After**, and although much of the story of stage carriage services at the turn of the century seems to have been written without the benefit of documents which have since come to light, it captures nevertheless the spirit and flavour of the period in a manner which only personal reminiscence can express. It therefore has much to lend to the history of Southdown in that area, the cradle of the Company. Ironically, it was the reluctance of James Town, one of the major contributors of those reminiscences, to turn from horse to motor traction which provided the opening for the launching of the Sussex Motor Road Car Co Ltd.

As in Brighton and elsewhere, the provision of horse-drawn omnibuses seems to have commenced with the coming of the railway. Numerous small proprietors came and went throughout the next half century, including Jinks, Norris, Scovell, Townley who ran between Worthing and Storrington (later covered by SMRC), and John Stent & Son whose services included much of Worthing to its eastern and western boundaries, together with a joint station-bus in conjunction with James Town.

Although well-established long before, from 1888 Town was running half-hourly headways from Heene, the Pier and the Post Office to Broadwater, and from the Post Office to Heene, latterly at a fare of 4d. (1½p) for the whole journey as a result of motorised opposition. Earlier, his slogan printed upon his tickets 'James Town's excursion coaches are the Safest and Best' - they ran to Bramber Castle, Arundel and Littlehampton - took something of a knock. In 1890 two horses attached to one of his service buses, fortunately empty and unmanned, bolted from the Town Hall to the Sea House Hotel where they attempted to pass each side of the piers of the hotel portico. The heavy porch was immediately swept to the ground, killing both horses and smashing the bus to matchwood. However, such was the sympathy felt for Town, his fellow townsmen subscribed a considerable sum toward the cost of the damage. Meanwhile, with nothing quite so dramatic befalling them, Jay's buses ran a quarter-hourly service between the Thomas a'Becket and East Worthing and Selden road to the Tarring Fig Gardens for 2d (1p).

In most of the larger municipalities at the turn of the century, electric traction was very much the up and coming motive power. In 1903, both Worthing and its horse bus proprietors survived a British Electric Traction attempt to build an electic tramway in the resort. Brought up with horses and quite unable to bring himself to purchase a motorbus, James Town turned a deaf ear to numerous pleas to try the new form of transport and send some regular motorbus services out into the surrounding countryside. Frustrated, the council commissioned a report upon the feasibility of organising its own municipal motorbus department, but eventually settled for announcing its preparedness to grant licences to "the several persons who have been making enquiries on the subject".

Sussex Motor Road Car Co Ltd

Accordingly a completely new syndicate, formed in the summer of 1904, received notification that their plans to link up the railway stations of Worthing and Pulborough would be greeted favourably. Beville Molesworth St. Aubyn of Toat Farm, Pulborough, whose idea the project was, had been in Brighton the previous December, when the Brighton, Hove & Preston United Omnibus Company's first motorbus rolled into town. Greatly impressed, he had seen in his mind's eye similar vehicles free-ranging in the Worthing area. St. Aubyn thus became managing director of the Sussex Motor Road Car Co Ltd, with his stepson Sir Walter Barttelot of Stopham as chairman, Newland Tompkins, as assistant managing director and joint-secretary. The other directors were Douglas B. Hall and James M. M. Erskine. The managing director's biggest achievement, however, was to convince Walter

The Sussex Motor Road Car Co Ltd favoured the Clarkson steam bus for its initial route linking Worthing and Pulborough, BP 319 being one of the two vehicles which entered service in November 1904. Boiler trouble caused by hard Sussex water led to their withdrawal and sale within six months, and they had a more successful second career in North Wales.

Flexman French, resident in his summer address at Telville Road, Worthing, that he should become manager and engineer of the fleet.

When French took up his appointment, the syndicate had purchased two Clarkson 'Chelmsford' steam buses to operate the service which commenced in November 1904. They were parked overnight in Shelvey's Mineral Water Works in Station Road, Worthing, until a garage was rented in the yard behind the Railway Hotel. Facilities comprised two pits, a workbench nine and a half feet by two and half feet, a ticket office the size of a single bedroom and, if the company buses returned before the cars of the hotel guests, a parking lot outside.

After the first few weeks of working, it was found necessary to install large underground tanks to collect rain water for the steam buses at the White Horse Hotel in Storrington, because the hard water of Worthing furred up the boilers which frequently burned out their tubes. The injectors also gave trouble and caused drivers to steer one handed and pump vigorously with the other. French soon realised that chalky Sussex water and steam engines were simply not compatible and convinced the directors that they should be sold whilst still in reasonably good condition. They went to the Vale of Llangollen Engineering Bus and Garage Co Ltd after less than six months' service.

In March 1905, a Milnes-Daimler 20hp saloon omnibus arrived to relieve the steamers and further buses of the same make were added for the summer season which was considerably helped by traffic to the Sussex Territorials camp at Washington. By the end of the year, four Milnes-Daimlers had been fitted with open-top double-decked bodywork to French's design. There were seats for two passengers beside the driver and a further cross-seat for four immediately behind him, all of which were preferred, even to those on top, by the travelling public despite a 50 per cent surcharge. The remainder of the seats downstairs faced inward. Rather than engage the owners of numerous overhanging trees in litigation, the seats on the upper deck were arranged to minimise the danger to passengers: the first three vehicles with back-to-back knifeboard seating and the other with five garden seats placed down the centre. The others were altered to this configuration by the spring of 1906, by which time the saloon had become a double-decker. Much of the clerical work was undertaken by the Pulborough Estate Office of Newland Tompkins and the company's good relations with licensed victuallers along the route was unquestionable. Its parcel offices and waiting rooms were established at **The Railway Hotel,** Worthing, **The Gun,** Findon, **The Maltster's Arms,** Broadwater, **Frankland Arms,** Washington, **The Half Moon,** Storrington, **The Crown,** Cootham, **The White Horse,** Marehill, and the terminus just beyond the railway arch in Pulborough, **The Railway Hotel.** Despite the fact that the drivers were not permitted to imbibe, the

SMRC switched to the Milnes-Daimler motor bus as a more practical proposition and four were in service with double-deck bodywork by the end of 1905. Walter Flexman French designed the bodies which, although at first glance were orthodox for the period, had several unusual features. The 'front' bulkhead was almost half-way along the structure to allow a row of seats to be fitted immediately behind the driver; upstairs the partially back-to-back seating arrangement can be seen. CD 338 was to be one of the oldest vehicles to be transferred to Southdown in 1915, though by then carrying a charabanc body. It is seen outside the SMRC office in Pulborough.

journeys were not without incident: on Wiggenholt Common a bus lost all its windows against a telegraph pole whilst avoiding a startled horse and another parted company with one of its wooden wheels whilst descending the Bostal—both without injury to their occupants.

Wooden wheels gave considerable trouble in the early days of motor traction. They were fine on horse-buses, where the tractive effort was provided by some outside force pulling at the bodywork they merely supported, but once the drive was by a live axle through the centre of the wheels, the spokes gradually leaned away from their proper positions, thus loosening the entire structure and adding greatly to the noise of the vehicles' progress. The effect was at its worst in dry spells and the SMRC was not without its pond to soak the wheels on such occasions.

In September 1905, SMRC started a second service between Worthing, Littlehampton and Arundel which did its best to provide connections with the Pulborough buses. The fare for that journey was one shilling and fourpence (6½p) and an extra sixpence (2½p) if the passenger wished to sit next to the driver unless, in his subjective judgement, she was young and pretty. SMRC also tried to begin running to Brighton in competition with another company founded in 1904, the Worthing Motor Omnibus Co Ltd. WMOC had been operating local services from Ham Road to Tarring Fig Gardens; Broadwater to West Worthing and West Tarring to the Town Hall, in competition with Town's and Jay's horse buses, with three 24hp Milnes-Daimler open-topped double-deckers which were actually on the road a few days before SMRC's 20hp Milnes-Daimlers. It added two more, all with 36-seat bodywork which was considered at the time to be rather too large for country work and in the absence of licences from Brighton, established a service to Palmeira Square, Hove, in June 1905. James Erskine, however, was a director of both companies: by September rivalry was averted and the SMRC buses merely provided connections for the Hove buses.

Soon after the directors of WMOC realised that a division of interest was counter-productive and, in August 1905, the Worthing Motor Omnibus Company was absorbed by the Sussex Motor Road Car Co Ltd. Worthing Motor Omnibus Company founders David Brazier and 'Major' Harry Gates sold their interest, but Secretary Arthur Stubbs continued to act in that role for his new company. B. Molesworth St. Aubyn remained managing director and Walter Flexman French was retained as manager and engineer of a fleet now doubled to ten vehicles. A major asset gained was the WMOC's partially constructed Ivy Arch garage on the Broadwater road, the other side of the railway from SMRC's original depot and potentially twice its size. The working environment of the Ivy Arch premises was already a great improvement on the Railway Hotel garage, from the channels and raised walkways of blue setts which kept the bus washers and crews' feet dry to the provision of a 12ft. by 15ft. workshop. Its equipment included a 6hp National gas engine, which supplied power and electric light for the building, and enough tools to maintain 25 buses—a clear indication of the expansionist intent of the management. The petrol store was licensed to hold 4,000 gallons which were delivered in 40-gallon steel barrels and mounted on stillages with, optimistically, a New Era fire extinguisher nearby—and to the consternation of the local petroleum inspector, there were 182 more gallons of petrol and 25 of benzolene stored illegally in a greenhouse outside.

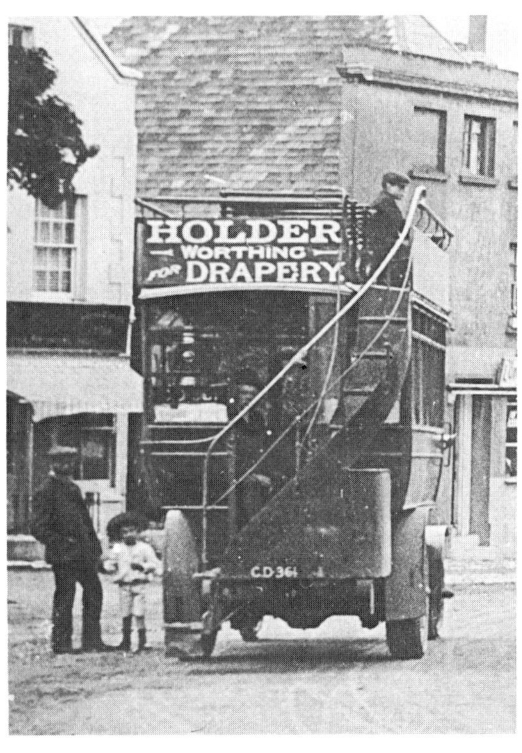

This rear view of another of the Sussex Motor Road Car Milnes-Daimlers, CD 361, shows the centrally-placed forward-facing seats used on the upper deck. This vehicle was also a survivor into the Southdown era, though it had received a slipper-type single-deck body after passing into Worthing Motor Services ownership and, like most of the earliest buses, was withdrawn by 1917.

Immediately after the merger, the clerical staff was relocated in French's house near the Railway Hotel and it was his policy to train all new drivers himself. The Company now planned to move farther afield and, in October 1905, introduced a new service between Pulborough, Steyning and Hove. There was also an over-ambitious plan to operate a Pulborough-Billinghust-Haslemere-Godalming-Guilford route with six additional buses but clearly the management realised that such a service would have been too long and the terrain too difficult. Instead, SMRC's renewed efforts to gain access to Brighton were now successful and the necessary licences were issued in January 1906. The Brighton, Hove & Preston United Omnibus Co Ltd actually pipped it to the post, however, gaining running rights over the entire Brighton to Worthing route on 18th December 1905. Brighton Corporation declined to give any reason why they had previously refused to grant licenses to SMRC. Said the journal **Commercial Motor** "it is ... probable that the extreme jealousy between Brighton and Hove has been allowed to eclipse all considerations of the public welfare". In fact, if one employs the Namier technique of the historian and looks at not the institution but the people involved, the membership of the board of directors of BH&PUOC tells its own tale.

Perhaps it was SMRC's chosen parking place in Brighton—an undertaker's yard—which presaged disaster. Mechanical troubles with the vehicles were numerous. For instance, the early Milnes-Daimlers were lubricated by an elaborated system of 30 sight-feed bottles of oil on the dash-board with copper tubes leading from each to the part to be lubricated. These the driver, in addition to his main task, had to keep dripping at the correct number of drops per minute and each one at a different rate. French removed the bottles and adopted greasers for most places, with two half-pint bottles for the crank case. Each driver then had to carry in the bus a quart filler-can and two gallons of oil to keep the crank case bottles topped up. Soon, footboards were covered in oil and the driver got his hands and clothes coated in a mixture of oil and dust. If earlier enthusiasm in the ranks waned somewhat it is hardly surprising. Lubrication problems, gear box trouble and difficulty keeping the tyres on began to accumulate and the crash of the **Vanguard** Milnes-Daimler on Handcross Hill came at a bad time for public confidence. The route to Brighton was not exactly what SMRC had hoped for either: it had wished to terminate the service at the Aquarium, but conditions imposed by the Council insisted that the buses should go on across the unremunerative part of town to Chesham Road, Kemp Town, thus adding unwanted mileage to a route on which it was already obliged to compete with BH&PUOC and the Brighton & Shoreham tramway. Daily breakdowns became more numerous.

In the summer of 1906, Walter Flexman French returned to London and entered the commercial car-hire business. SMRC struggled on with the vehicles under the supervision of engineer A. E. Hartley, until by the early spring of 1907 it was making a heavy loss on the Worthing-Brighton run. Engaged a few months earlier as consulting engineer, Alfred Douglas Mackenzie now accepted the position of general manager of

SMRC with his assistant Cannon to act as a very youthful chief-engineer, thus setting in motion what was to prove a thirty-seven-year partnership.

Alfred Cannon immediately set about rationalising the maintenance system and Mackenzie attempted the bold throw—expansion, and far afield at that. Remembering Bartlett, he headed for Portsmouth and on Saturday 12th October 1907, Milnes-Daimler CD 408 rolled to a halt at the Commercial Road office on a trip from Bognor. Three journeys per day were then run via Chichester, Emsworth and Havant. It was planned to close the gap between Bognor and Littlehampton the following spring, thereby establishing a Brighton-Worthing-Littlehampton-Bognor-Portsmouth facility serving all the towns and villages along the line of route. This grand strategy was thwarted, however, by sheer lack of custom at the Portsmouth approaches. On 6th December 1907, Mackenzie obtained permission from Portsmouth Council to transfer two of the four buses used on this route to run within the town itself; from Edinburgh Road to Eastney Barracks via Arundel Street and Fratton Bridge,; and from Clarence Pier, where there was an additional booking office, to Eastney via South Parade the following spring. SMRC rented a garage in Rudmore Lane to house the vehicles. That winter an additional bus clawed its way over Portsdown Hill on a route to Hambledon, the birthplace of cricket, and spent three days stranded in a snow-filled ditch. 'Last bus' crews slept overnight at the 'George' inn at Hambledon and the bus was parked in the yard, an arrangement which led to a petrified cockerel taking an unnoticed journey to Portsmouth perched on the back axle, whence he was retrieved by the crew and returned to his hens in a sack.

Mackenzie's plans to save the Sussex Motor Road Car Co Ltd included the purchase of an MOC and two Thornycroft coaches fitted with 20-seat charabanc bodies (originally built for the Milnes-Daimlers) so that body-swaps could provide for seasonal change. SMRC vehicles appeared at Goodwood and Ascot Races, and in addition to the office which Mackenzie retained at Ryde, offices were opened for excursion traffic at Seaford, Newhaven, Eastbourne, Hastings and at the 'George Inn' yard near the Bargate, Southampton. Southampton Corporation was at the time considering restarting a motorbus service—which it first ran in 1900 and 1901—between the Clock Tower and Northam, and Mackenzie tried to secure it for the SMRC. If he had, it is highly unlikely that Hants & Dorset Motor Services would ever have been founded and Southdown, one may speculate, would eventually have provided stage carriage services all the way from Hastings to Bournemouth. Instead, Harry Lawrence drove a coach, without the benefit of a Southampton licence, from the 'George Inn' yard to Lyndhurst and back for a couple of months in the summer of 1908. If the local police didn't notice, the station master at Totton certainly did, for he was so impressed with the vehicle which paused there upon each trip, he arranged for the London & South Western Railway company to engage Lawrence as the local vehicle mechanic, whence he eventually returned to spend his working life driving for Southdown.

The three new coaches were fitted with what was described as 'slipper-type' bodywork to Mackenzie's own design. Six rows of seats, each five inches higher tier above tier, with two more beside the driver gave 32 passengers a perfect view of the road. Registered DL 261, DL 208-9 respectively, they were almost certainly ordered by Mackenzie for the Isle of Wight Express Motor Syndicate Ltd, but redirected by him to SMRC. Initially, the MOC worked the Worthing-Brighton service and the Thornycrofts were employed at Bognor and Seaford, their curious shape arousing great interest wherever they went. Meanwhile, the Hampshire adventure of SMRC ended in the autumn of 1908, when the four vehicles based in Portsmouth were hastily withdrawn to Worthing because there was not enough money to pay the garage rent. Things were not much better in Worthing itself either: despite the fact that on one trip Bill West found himself driving Queen Alexandra and the Czarina of Russia from Brighton to Worthing, it became necessary for some additional running along Marine Parade in the evenings to get enough money to pay the wages. In November 1908, the official receiver for Brighton was called in.

Renowned for his reluctance to throw anything away, Douglas Mackenzie had, the previous year, purchased from the London Standard Motor Omnibus Company six of the original Milnes-Daimlers of the defunct Hastings & St. Leonards Omnibus Company and stored them about the Ivy Arch garage during the attempted expansion. Since he'd paid only £90 for the lot and "had difficulty in getting one serviceable vehicle out of the assembled pieces", it is possible they were initially intended merely as spare parts. Nevertheless, because the SMRC had not enough funds to purchase them from Mackenzie, the receiver now declared them to be his property. Following the liquidation of SMRC, Mackenzie and Cannon breathed life into two of these buses and, keeping their band of employees together, ran them between Worthing and Storrington throughout the winter of 1908-9.

Formation of Worthing Motor Services

Douglas Mackenzie had been taking out metaphorical insurance in another way also. He had retained his office in Victoria Street, Westminster, and in May 1908 this became the registered office of two allied companies, of which Mackenzie was again the consulting engineer, Western Motor Coaches Ltd of Minehead, Somerset, and Kent Motor Services Ltd of Maidstone. KMS had been started, at the request of local residents who assured Mackenzie that there was ample traffic, to run between Maidstone and Sutton Valence, Sittingbourne and Faversham, and from Maidstone to Hastings on Sundays. In fact the Kentish enterprise lasted little longer than two months and its two Milnes-Daimlers and two Thornycrofts were sent along to Worthing.

The arrival of Douglas Mackenzie as General Manager resulted in some fresh ideas on vehicle design. This Thornycroft, believed to be an 80B4 model and having a similar radiator and bonnet to those used on Thornycroft private cars of the time, was delivered in July 1907, being the first of two placed in service by SMRC. The body was similar in its tiered-seating principle to one built to Mackenzie design on a Leyland X-type for a Clacton operator the previous August.

The third of the trio of new vehicles added to the fleet was this MOC, registered DL 261, and seen on the Worthing-Brighton service it operated from April 1908. This time the bodywork was to the Mackenzie slipper design in nearer its original form, complete with roof and, in this case, railway-style mouldings on the side panels. Route details, including starting time, were shown on boards carried at the bottom of the sloping sides. The MOC was built by Motor Omnibus Construction Ltd, a subsidiary of London Motor Omnibus Co, better known as 'Vanguard', based at Walthamstow, using parts supplied by Armstrong Whitworth, production ceasing as a result of mergers about this time. What is thought to be another view of the same body, in WMS livery, below, shows the extraordinary appearance of the design even more dramatically. The reference to 'coaches' in the wording about the Brighton departure point is noteworthy—this use of the term for a motor vehicle was not to become common practice until nearly 20 years later. This time it was on a Milnes-Daimler.

The way in which bodywork was regarded as a matter for variation is shown by these two views of Milnes-Daimler CD 393 in the Worthing Motor Services Ltd fleet. In the scene above, it has a 'rudimentary' slipper-type body, similar to that on the Thornycroft shown on the opposite page. The canvas side screens are reminiscent of the practice on contemporary paddle-steamers. The same vehicle is seen on the left with a much better-finished body with top cover to eliminate the need for ladies to carry parasols evident above. The high mounting of the registration plate, on the front of the canopy, was to remain a Southdown characteristic until the 'fifties.

Formed in March 1909 to acquire the Ivy Arch garage and work the remaining vehicles of both the old SMRC and Douglas Mackenzie, Worthing Motor Services Ltd was registered the following month. In order to keep the registered numbers in sequence, the methodical Mackenzie had used his Ryde office to provide DL series numbers for the ex-Hastings chassis he'd saved from the bits and pieces. The new company retained Mackenzie and Cannon in their respective roles and then afforded them the status of directors. It extended the Worthing-Storrington route back to Pulborough, and plunged into the private hire market in May 1909, providing the transport for the Sussex Royal Garrison Artillery's 'Motor Dash' from Brighton to Newhaven Fort—a journey accomplished in 30 minutes, with Milnes-Daimler double-decker DL 383 well to the fore. Leaving Portsmouth alone for the time being, WMS lost no time in applying for new licences to run into Brighton, and in restarting the local services at Worthing, and to Arundel and Littlehampton—with a total of fifteen buses. Additionally, it acquired rented premises at 23 Marine Parade, Worthing, and into this, from 'up the stairs' at the Ivy Arch, was moved the administrative, clerical and miscellaneous duties section of the company. In this building a clerical assistant spent a laborious week stamping new fares on old Isle of Wight Express Motor Syndicate tickets for use locally and then, when Mackenzie had a better idea, glued paper panels over the Isle of Wight name and destinations upon each ticket and wrote in the Sussex ones instead.

Another transformation is seen in these two views of Milnes-Daimler DL 382, (above) with double-deck body and bearing route boards for the Worthing-Storrington service, though no fleetname, and (left) with a later development of the slipper type of body, having full-drop windows in the body sides. The full load in the high-mounted rear seats cannot have helped stability.

Such economy gives a clear indication of the financial state of Worthing Motor Services Ltd in its first four years and, indeed, it was not until 1912—when Alfred Cannon himself purchased a Straker Squire chassis and hired it to WMS—that the company managed to add to the fleet. Meanwhile, WMS' entry into Brighton was the subject of new council restrictions. Previously the licences had been granted on a protective-fare basis: now there was to be no plying for hire on the Brighton side of Portslade. Worthing town council retaliated by imposing similar restrictions upon the Brighton Hove & Preston United, leaving each company to bring back only the passengers they had taken out—which is hardly what a stage-carriage service is supposed to be.

Each company now circumvented the restriction by opening an office in the other's home town and getting the passengers to book in advance. BH&PUOC rented No. 5 South Street, Worthing, and WMS came to an agreement with Stuart Smith to make use of part of his Royal Mews in Steine Street, Brighton. The remainder of those premises was occupied by cabs and their drivers, who viewed motorbuses with undisguised disfavour. One of them lost no time reporting George Cowley to the council authorities when he himself went to the office to buy the distinctive square tickets for the passengers, instead of sending them in to make the purchase. The technique employed by the drivers and which he should have followed was to drive slowly round the Aquarium to attract attention and lead the customers back to the office. In effect, Stuart Smith was WMS manager in Brighton.

Prior to the January 1912 agreement with BH&PUOC which eliminated the need for such antics, it was WMS policy to require all passengers to book at the office, even in Worthing, where it was actually unnecessary, to ensure that they did not travel on the rival buses. Both companies took to wearing flags to identify themselves to the faithful; BH&PUOC a Union Jack, and Worthing Motor Services Ltd a red pennant bearing its initials. It was during this period that driver Wally Turner, who was to retire from Southdown in 1950, earned his nickname 'Frottle'. "Get down on that Frottle" he would say to his conductor when he wanted his colleague to reach under the bonnet to force down the throttle rod and help him outrun the BH&PUOC bus to the next stop.

After 1912, the vehicles released from the competition of the Worthing and Brighton road were able to go in search of new custom elsewhere. The Worthing company's buses now ran from Brighton to Newhaven and Seaford without any attempt at opposition from BH&PUOC, in return—on a reciprocal basis—for five per cent of WMS' takings on that route to be paid to the Brighton firm. Those fares which were subject to this commission were taken in exchange for an unmistakeable series of yellow tickets. Fully *au fait* with railway history, and no doubt mindful of the time when the London & Southampton railway found it necessary to change its name to the London & South Western, in case it upset the people of Portsmouth whose town it was about to serve, Mackenzie now decided that the 'Worthing Motor Services' label was a shade too parochial for coaches about to operate in Brighton and as far afield as Seaford, and the name 'Sussex Tourist Coaches' was adopted for excursion work and coaches working east of Brighton.

It was during this period that Mackenzie began to work with the flair which established him as a traffic expert of considerable repute. He had the knack of driving a bus at its lowest oil-consumption, and expected others to follow suit, of knowing by the slightest noise that a vehicle needed attention, of knowing when to change the role in which a vehicle operated. The man whose intention in life was "to get them off their bicycles" would arrive at the Ivy Arch garage as fast as his own bicycle would carry him and call upon everyone in sight to drop whatever he was doing and give a hand with 'body-shifting'—changing coaches to buses and vice-versa— in order to get numbers and usage in their proper sequence. There was nevertheless a great cameraderie at the Ivy Arch garage and that famous 'Southdown' style came another step along the way. Among those who 'fielded' the pay packets which the young Alfred Cannon 'handed' out was Charlie

The Worthing Motor Services offices at 23 Marine Parade, Worthing. The board shows a dozen departures for Brighton (giving an hourly service, or better, from 10.15am to 7.15pm), four for Storrington, three for Arundel via Littlehampton, two for Long Furlong and two on the 'Goring 6d (2 ½ p) Circle' as well as single timings for Bramber and Arundel via Amberley.

Here the slipper body has progressed to a fully-enclosed vehicle with glazed front bulkhead, though almost every window would open. The curved line of the skirt panel, a logical development of the styles shown opposite and on page 7, gives quite a rakish effect and a detail feature anticipating much later practice is the use of inclined glass louvres to allow some of the side windows to be opened slightly in wet weather. The Sussex Tourist Coaches fleetname puts the date at not later than 1912—the chassis is reputed to have been a Leyland X-type, the radiator being of a fluted style used by Leyland on some chassis around 1906, but other chassis details do not support this.

This 1913 Leyland X-type 4-ton model was among the Worthing Motor Services vehicles which passed to the new Southdown company in 1915, surviving until 1921, and thus playing a key part in setting the pattern of "never being without a Leyland" true of the Company to this day. Its registration number, DL 493, showed that the Mackenzie connection with the Isle of Wight was still active. The Dodson body seated 39. Note how lettering and livery was nearing early Southdown practice.

The last additions to the WMS fleet were Daimlers. The products of the Coventry-based Daimler company had become as widely respected among British operators in the 1913-14 period as had been their earlier namesakes of German origin, with which all connection had long been broken. This CC model with 36-seat Dodson double-deck body was registered IB 707, registrations of WMS and subsequently Southdown vehicles often having this Irish (County Armagh) mark in the period up to 1919, this being because of willingness to issue numbers in blocks with final digits matching those of the fleet numbers, in characteristic Mackenzie manner. Many Daimlers were impressed for military use in 1914 but this chassis continued in service as Southdown No. 7 until 1920. It had been re-registered CD 5007 in 1919. The body had a complex series of transfers, becoming the 'milk churn' body shown on page 24.

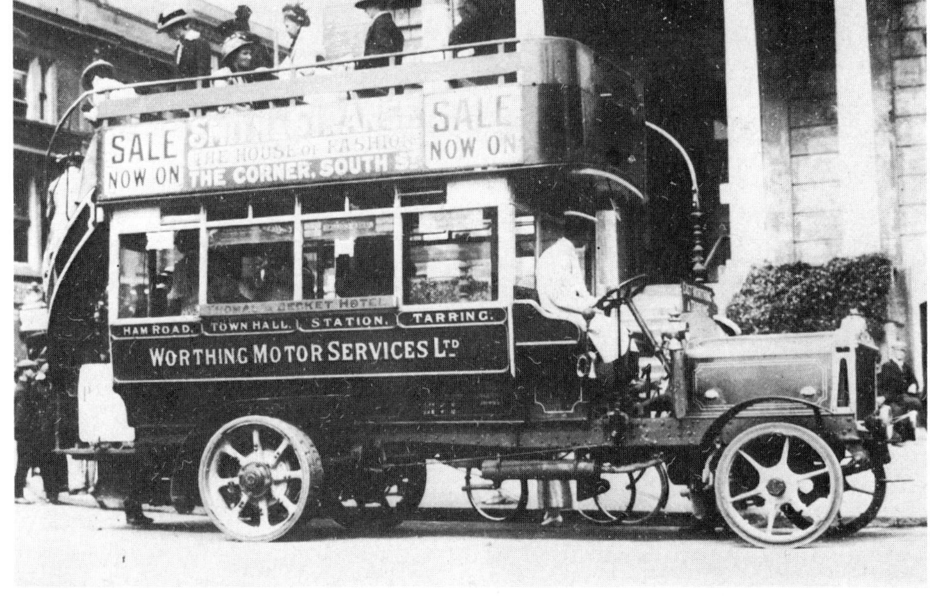

Harris, the 'car' washer, William Treby who burnished the big ends, driver Len Pearson with his immaculate leggings, Charles Gates, Bill Ellis the leading tipster, R. F. Clement who was to be awarded the MBE in 1953, H. E. Humphrey who became garage superintendent at Littlehampton, W. J. Cooper—a conductor who became area manager at Worthing, Bill Turner—an ex-Brighton horse-bus conductor, and little Jimmy Tee and Scott Fairbanks who were soon to go off to Flanders - and never return. Others were to spend their entire working lives with Southdown. George Cowley who was to transfer to Portsmouth and was always known by Cannon as 'Young Cowley' retired in 1959 with a record of 51 years in PSV driving. The very last to retire was to be conductor William Jay, whose father's horse-bus services in Worthing were taken over by WMS in March 1913.

This was to be the year when Worthing Motor Services began to purchase new vehicles. A pair of Tilling-Stevens double-deckers joined the Worthing fleet which was now operating the following departures each day: Brighton 12, Findon, Washington and Storrington 4, Littlehampton and Arundel 3, Goring Circular 2, Findon and Bramber 1, Amberley and Arundel 1. More importantly, the Brighton, Newhaven and Seaford excursions were greatly enhanced by the purchase of Daimler vehicles for the 'Sussex Tourist Coaches' fleet. Advertised on bill-heads and tickets with the slogan 'Green Cars run Everywhere', Douglas Mackenzie was about to realise his ambition to show 'Beautiful Britain' to a completely new kind of tourist (see chapter 6).

3: London & South Coast Haulage Co Ltd

No sooner had Worthing Motor Services and the Brighton, Hove & Preston United Omnibus Company reached their 1912 accommodation and subsequent *modus vivendi* than a new company and potential competitor arrived in Brighton. Having left Worthing in 1906, Walter Flexman French had made yet another mark for himself as London manager and agent for the Ryknield Motor Co Ltd, launching French's Ltd (Motor Jobmasters) and hiring out Ryknield lorries on a contract hire basis. This firm became French's Garage & Motor Works Ltd which apart from carrying out the business which its name suggests, became the main agency through which vehicles for Maidstone & District and the future Hants & Dorset Motor Services were acquired and the wages of their employees were paid in their formative days. French, together with H. E. Hickmott, better known as Major Hickmott with Ribble Motor Services Ltd at Preston, in Lancashire, in the years between the wars, now took an interest in the London & South Coast Haulage Co Ltd, which set up in Brighton with two lorries and a van.

London & South Coast Haulage undertook contract work and cartage for a number of local firms including Hannington's the furniture stores. Within two years, the venture was in danger of imminent collapse as neither the amount of work nor the rates obtainable permitted the company to break even. Turning to a sideline in which he had already enjoyed mixed fortunes, French decided that passenger transport might prove the saving of their faltering enterprise. In Kent, he had already formed omnibus companies at Margate and at Maidstone, his son George now ran the fledgling Maidstone & District Company. In June 1914, French purchased a Daimler CD saloon motorbus and applied for a licence to run from Brighton to Hurstpierpoint. Following an initial rebuff, this was granted after a free excursion followed by tea had convinced the members of the Watch Committee that they could hardly refuse.

To reinforce their entry into the bus field, L&SCH purchased as a going concern the business of William and Anne Ecclestone, whose 'Brighton Queen' charabanc operation was better known as 'Jolly Jumbo's' - a reference to Mr Ecclestone's outsize dimensions. The Ecclestones had provided horse-drawn excursion vehicles from their leasehold Middle Street premises and only recently converted to a fleet of five Fiat touring cars and charabancs which they ran from the Kings Head Garage in West Street. At the same time, L&SCH purchased four torpedo-bodied Durham-Churchill charabancs and four further Daimler saloon buses, extending the Hurstpierpoint service to Burgess Hill and opening a second route to Falmer and Lewes. The stage carriage buses bore the somewhat misleading title 'Brighton and South Coast', for its routes went inland.

At the sharp end of the London & South Coast Haulage company's affairs were Driver E. W. Parsons who was to undertake some pioneer work for Southdown, and would be the last of the L&SCH men to retire from its ranks in 1950, Richard Ried conductor and driver, Cecil Pullen, Bill Holland, Frederick Kerman who became garage foreman at Freshfield Road and R. C. Paramor who was Southdown's mechanical inspector at Bognor in the early 'fifties. The best efforts of these men presented a considerable problem to the Brighton Hove & Preston United Omnibus Co Ltd who had themselves sought licences for the Lewes road. Its difficulty was compounded when the L&SCH added a Straker Squire and the Leyland S-type saloons to their fleet and intensified their headways, operating from both of the ex-Ecclestone properties, with additional storage facilities at Ditchling Rise. These were the days when a new saloon omnibus, chassis, tyres and body complete, cost in the region of £800. Two of the L&SCH stage carriage vehicles were secondhand. The Daimlers were particularly modern and attractive to customers and were some of the first buses to provide drivers with a windscreen to protect them from rain, flies, dust and sprayed tar. It was clear that with the kind of journey that French could now rustle up for his operations this, potentially, was a serious time indeed for the incumbents.

Just what kind of battle would have developed between Walter Flexman French and the man who had replaced him at Worthing in 1907 was never to be put to the test. After 24th August 1914, 'French' meant Sir John French or THE French, as the British Expeditionary force, tied by political considerations to the strategy of its ally, found itself retreating from Mons—the first battle of The Great War, as it became known until subsequent history renamed it World War I. Commercial 'battles' at home became positively anti-social overnight.

Walter Flexman French returned to bus operation in Sussex in 1914 when his London & South Coast Haulage Co Ltd began operations under the Brighton and South Coast fleetname. Bodywork of advanced design for the period was chosen for four Daimler buses, the driver of CD 2645 having both a windscreen and a cab enclosed at the side. The chassis was soon to be requisitioned for military use but the bodies passed to Southdown—see opposite page.

Among the new Southdown company's inheritance from its predecessors was this Straker Squire U-type, CD 1681, No. 28, which had come from the Brighton, Hove & Preston United fleet, being the sister vehicle of CD 1441, shown on page 11, though in this case slightly newer, dating from 1912. Both chassis survived in the operational fleet until 1925, though several body transfers had by then taken place. The date of the photograph is not known, but the body shown appears either to be the Dodson original, carried until 1917, or that from CD 1441, carried between 1919 and 1923, in both cases by Dodson to a clearly 'slipper'-inspired design. More Straker Squire vehicles, new or second-hand, were purchased in the period up to 1918.

Chapter Three: The founding of Southdown

If one employs the maxim about silver linings, it could be said that but for World War I, Southdown Motor Services Ltd might never have come into existence. At one and the same time the outbreak of the conflict presented the three companies concerned with both a sudden increase of people on the move and a government policy-decision which constrained them from capitalising upon the situation. In the early days, it was not so much the employees who enlisted in the forces, as the British Army's need to bolster its transport columns which created the necessary climate for drastic rationalization.

As early as 1904 it had been recognised by the military authorities that motorised transport would play an important part in future wars. The notion that new officers should be given instruction in engineering was given serious consideration. Several years later Worthing Motor Services' participation in the Territorials' 'Motor Dash' to Newhaven marked one of the earliest attempts to move large bodies of men by motor vehicles. In 1911 it was estimated by the War Office that, in the event of a European war, the British Army would require some 900 lorries. Rather than put the Army Service Corps to the trouble and expense of maintaining that many vehicles in peacetime, a subsidy scheme was introduced whereby heavy goods—and for that matter passenger—chassis could be acquired at a reduced rate if the operator purchased those built to a specification prescribed as rugged enough to withstand conditions of the field. Prominent features of such vehicles were the relatively high ground clearance and sturdy dumb-irons to which towing hooks could be firmly attached. Among the twelve main manufacturers of these vehicles, Daimler of Coventry was well to the fore—and all three Sussex companies had invested in the make.

Just as, in the Boer War, the best horses had gone from the BH&PUOC stables, army requisition officers now began to visit the depots and arrange for the removal of the better chassis. By the end of 1914, BH&PUOC had fourteen spare bus-bodies, WMS ten and L&SCH five: all of the latter and most of the others marked the passage of army Daimler-collectors. The Army Service Corps entered the war with 950 lorries—and finished in 1918 with 33,500. Clearly what was taken from Brighton and Worthing was but a drop in the national bucket, but the effect locally was immediate. The companies put aside their rivalry, helped each other out with the loan of surviving vehicles and began to talk about ways of resolving their common problem. Amalgamation, it was decided, was the answer.

For all intents and purposes, save the legal niceties, the decision to form one new company had been worked into some detail by the end of 1914—and it began operating

The new Southdown company, like many other operators, had to turn to unfamiliar sources for chassis amid wartime shortages. Among them was Caledon, made by Scottish Commercial Cars Ltd of Glasgow, a former Commer dealer unable to obtain supplies of that make of vehicle. The resulting 4-ton model had a Dorman engine. CD 3326 was one of three purchased by Southdown and fitted with bodies acquired with the London and South Coast Haulage fleet which had originally been on Daimler chassis taken over by the military. This one, by Brush, was probably from the bus seen on the opposite page. It is seen in Lewes in October 1915, soon after entering service; within a year all three chassis were sold.

as such with effect from 1st January 1915. Among those involved in the planning, it was Walter Flexman French who carried the greatest 'clout'. Already chairman of Maidstone & District Motor Services Ltd, he had succeeded in getting the mighty British Electric Traction Co Ltd to take an interest in that firm and Sidney Garke and William Wreathall had joined the board on BET's behalf in June 1913. To BET, who had already invested in the purchase of the now defunct Shoreham Tramway, and singed their fingers as a result, this was the opportunity to repair their relatively mild embarrassment in some style. William Wreathall would now join the board of the proposed company to represent BET and in return for this vital introduction French was to be made chairman of the directors. Alderman John J. Clark and Sir John Bradford from BH&PUOC and Mackenzie and Cannon from Worthing Motor Services Ltd would complete the board. Between them they raised an initial capital of £51,250. At 31, Alfred Cannon became managing director and Douglas Mackenzie chose to be traffic manager. Frank Smith from BH&PUOC was secretary, his firm giving up its office in Worthing and subscribing to the pool its Steyne garage, Brighton; land at Freshfield Road, Brighton, and stables and workshops in Upper St. James's Street. As its first, and as it turned out temporary, headquarters the new company was to acquire from the same source an office at 6 Pavilion Buildings. From Worthing Motor Services came a garage on the Amberley Road at Storrington and rented accommodation at Marine Parade, Worthing; the Old Brewery House, Newhaven; the Esplanade, Seaford, and the Royal Mews, Brighton. The London & South Coast Haulage Co Ltd contributed its garage in Middle Street, Brighton.

The Brighton, Hove & Preston United Omnibus Co Ltd was to continue running its town services in the two boroughs under the direction of their superintendent and engineer E. A. Eager (until 1916 when, following the purchase by Tilling he became chief engineer of the Aldershot & District Traction Co Ltd). Finances, who-was-to-do-what and premises sorted out, and with the final count of the vehicles very much in the lap of the War Office, the directors designate now began to think about a name for the company: the new firm was to be 'South Coast Motor Services Ltd'—and, as the articles of association were drawn up, this was the name upon the title page and in all correspondence relating to it. The name appeared on the advance publicity, letterheads, timetable booklets and even a sizeable batch of tickets to be issued on the cars. Messrs Emile M. Marx & Colbourne, solicitors of Brighton sent the memorandum and articles of association to the Registrar of Companies in London and company number 14035 was issued. The date of incorporation was entered as 31st May 1915—and then someone noticed; the name was not acceptable. It was considered too similar to that of the London & South Coast Motor Services (1915) Ltd, a very old-established firm based on Folkestone which had just been re-registered by its new owner, a well-known pioneer in the business, Percy Allen. With just two days to think about it, Clark and French came up with the name Southdown, hastily amended and initialled the document and returned it to London. Southdown Motor Services Ltd was incorporated on 2nd June 1915.

Among the 27 objects for which the company was established are the following: (b) to purchase, lease, establish, or otherwise acquire any tramway or light railway, and any trolley vehicle (being a mechanically propelled vehicle adapted for use upon roads and moved by electrical power transmitted thereto from some external source) motor or other omnibus, ship, boat, aeroplane or airship, or motor and other cab undertakings in the United Kingdom or elsewhere, and to construct any motor omnibus, motor or other cabs, trolley or other vehicles (whether earth, air, or water borne) tramways, or light railways, and to equip, maintain and work, either by wire rope, or by electrical power, or by any mechanical power whatsoever, all tramways, light railways, trolley vehicle, motor or other omnibus, ship, boat, aeroplane, or airship or motor or other cab undertakings belonging to or leased to the Company or over which the Company may possess a right to run ... and work. (d) To carry on in the United Kingdom of Great Britain and Ireland and elsewhere on the Continent of Europe and in any part of the world the business of Tramways, Light Railways, Trolley Vehicles, Motor or other Omnibus, Motor or other Cabs, Carriage and Van Proprietors, and generally of Carriers of Passengers, Mails, and Goods by earth, air,

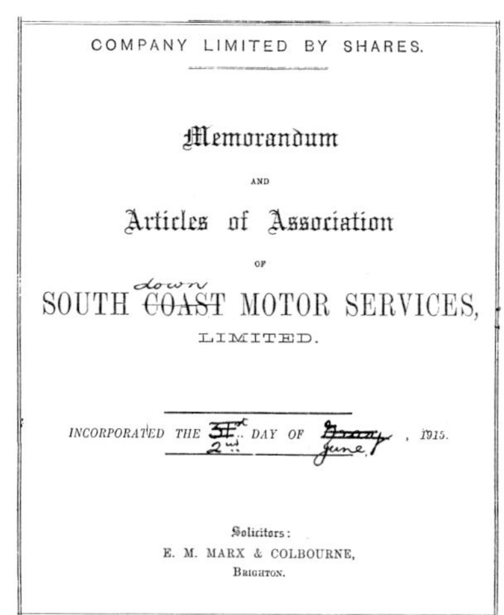
COMPANY LIMITED BY SHARES.

Memorandum

AND

Articles of Association

OF

SOUTH ~~COAST~~ down MOTOR SERVICES,

LIMITED.

INCORPORATED THE ~~31st~~ 2nd DAY OF ~~January~~ June, 1915.

Solicitors:
E. M. MARX & COLBOURNE,
BRIGHTON.

or water and of Manufacturers of and Dealers in Tramway and Light Railway Cars, Trucks, Omnibuses, Cabs, Trolley Vehicles, Carriages, Conveyances, Vans, Locomotives, Accumulators, Dynamos, and other chattels and effects required or suitable for the making, maintenance, equipment and working of tramways, light railways, or trolley vehicles, and for the carrying on of the business of carriers of passengers, goods, mails, and parcels by earth, air, or water, and of Omnibus, Carriage and Van Proprietors, Motor Garage and Show and Sale Room Keepers, and of all articles and things used in the manufacture, maintenance, and working thereof; and to carry on the business of Mechanical and Electrical Engineers, Machinists, Metallurgists, Saddlers, Galvanizers, and Packing Case Makers, or to manufacture or employ any person, persons, firm, company, or corporation to manufacture for the Company any motor cars, motor carriages, motor cabs, motor omnibuses, motor carts, cycles, bicycles, tricycles, or vehicles of all kinds whether earth, air, or water borne.

(The mark of Walter Flexman French is writ large in all this bet-hedging and is based upon his experience at Balham, Guildford and Maidstone. An almost word-for-word series of objects was used to launch Bournemouth & District Motor Services Ltd (Hants & Dorset), of which he was the prime mover, the following year.)

A glance at the map and the name seems obvious; the Southdown Hills provide the beautiful backbone of the territory the Company had its eye upon. Yet it is worth noting that Worthing Motor Services' new garage accommodation at Storrington, acquired as late as October 1914, was located in the westernmost portion of Adela Powell's 'Southdown Garage' and it is possible that it was this which provided the inauguration. Whatever the answer to that, the substitute name did not please all the directors. One at least complained that it reminded him of mutton and, if the nautical

This Daimler, 2-ton CB type, joined the fleet in May 1915, from Wilts & Dorset Motor Services Ltd, Mackenzie and Cannon's 'other' company, though it had originally entered service the previous year with Worthing Motor Services Ltd. It became fleet-number 3, rejoining numbers 1 and 2 based on similar chassis and registered IB 701 and 702 transferred directly from that fleet, though it is noteworthy that a series beginning at IB 801 was in use for Wilts & Dorset vehicles, as illustrated on page 22. The 27-seat charabanc body was by Harrington, a regular supplier until the 'sixties.

The first acquisition of another operator took place in May 1915, when Arthur Davies sold out his business in Bognor. With it came three Commer WP1 models with Bayley 27-seat bodies. The chassis were chain-driven and hence of less interest to the military than later Commer models. This earlier vehicle was called 'Lady Betty' but the registration of the last of the three, BK 2299, was probably enough to give Douglas Mackenzie the idea of numbering it 99 and the others 97 and 98.

Two McCurd chassis were among the collection of uncommon makes of new vehicles purchased in 1915, surviving in passenger service until 1919. The identity of this one is uncertain, but was probably No. 22 (CD 3322), which entered service with a Brighton, Hove & Preston United 36-seat rear-entrance bus body acquired as one of the spare units on the formation of Southdown. It was longer than usual for the period and the body in question was shortened before transfer to a new Leyland N-type chassis in 1919; it had been removed from the McCurd in 1916.

counterpart was anything to go by, a change of name was extremely unlucky. Not so, it turned out, with Southdown. It had from the beginning a well-balanced team of officers who were now to take it from strength to strength.

An impressive-on-paper total of 76 vehicles (and 19 horses and their harness) had been acquired. In fact, only 31 were considered suitable for service as buses, the horses had been valued at £17.10s.0d (£17.50) each and the remainder of the vehicles consisted of small charabancs, lorries, vans, taxis and all those horse-drawn coaches bestowed upon Southdown by BH&PUOC. The latter were pruned and actually kept on the strength, continuing to provide excursions from the Front at Brighton under the supervision of their former owner, F. J. Mantell. After the Emergency Provisions of 1916 whose purpose was to protect road surfaces by restricting new licences and the stricter rationing of petrol, the horse-drawn carriages proved an unexpected boon, much to the delight of Mantell, whose tickets to the Devil's Dyke got a new lease of life. So too did many other tickets. Under the waste-not-want-not Mackenzie, there was more sticking bits on Isle of Wight ones and those from all three constituent companies, including some of the old 'Sussex Motor Road Car' stock. Some of 'Jumbo' Ecclestone's 'Brighton Queen' charabanc tickets got issued on the excursions from the sea front as well—a case of 'those not skinning, hold a leg'.

The firm which thought it was South Coast Motor Services Ltd actually ran services with effect from 1st April 1915, and the following month had acquired a base in Bognor when it purchased Arthur Davies' charabanc business (see Chapter 4), placed him in charge as local manager and told him that wasn't who he was working for after all. His firm was of interest to Southdown not so much for its charabancs as the licence he had acquired in the spring of 1914 to run a stage carriage service from Bognor to Portsmouth via Chichester. Although not utilised for some while, this enabled Mackenzie to secure for Southdown a route he'd previously served. On its stage carriage routes the new company ran the ageing Milnes-Daimlers and Straker Squire, Daimler, Tilling-Stevens, and Leyland buses and even pressed into service Durham-Churchill charabancs— until Southdown began buying chassis makes of little interest to the War Office. Thus before 1915 was out Ensign, McCurd, Romar and Caledon were represented in the fleet in addition to more of the familiar makes both new and second-hand—a total of 25 additional vehicles, many of which now received the bodywork left spare by military acquisition.

In the summer of 1915, Southdown operated five routes from Brighton: to Worthing; Steyning; Lewes; Eastbourne and to Cuckfield, the in-town East and West Worthing route and one from Worthing to Storrington which, to mark its historical importance, remained thereafter Southdown's route No. 1. It also offered morning and afternoon charabanc excursions from Brighton, Worthing and Bognor. Meanwhile, keeping strictly within the borough, Tillings arrived in Hove and began the campaign which led to their taking over the remaining operations of the Brighton, Hove & Preston United Omnibus Co Ltd. Clearly, there was growing traffic in the area and, despite the earlier troubles and as long as the right makes were purchased, there seemed to be vehicles available. Within a year, French was able to write "... at Brighton there are 62 buses on outside service, independent of those strictly confined to the town ... at much less than railway fares".

Petrol supplies for Southdown were collected from the Silvertown depot of the Anglo-American Oil Company in 50 gallon steel barrels. The railways were still rather anxious about carrying such dangerous cargo and what had been a 20-seat coach was stripped to the floor and, dubbed the 'pig-boat' by its crew, carried out most of the fetching and carrying. Eventually, however, petrol rationing became so severe that several Scout and Caledon saloon buses

Another rare make to join the fleet was the Scout, built in Salisbury five of which were placed in service in 1914-5 by Wilts & Dorset Motor Services Ltd, being given consecutive registrations IB 801-805 to suit the fleet numbers 1 to 5 in typical Mackenzie style, and with the co-operation of Armagh County Council, block issues being rare at the time. The vehicle illustrated, new in May 1915, was one of four 37hp models and was sold to Southdown in March 1916 when it acquired the Dodson 30-seat body shown, the latter originally coming from one of the Brighton, Hove & Preston vehicles.

were converted to run on town-gas. George Stocker, who retired as Washington inspector in 1950, always claimed he was the first Worthing driver to take one out on service. The gas was stored in a rubber and canvas envelope fitted to the roof of the cars, whose service was restricted to a reasonable distance from the town supplies of Worthing and Brighton. As the vehicles reached the latter stages of each journey they took on a distinctly dishevelled appearance as the near-empty bag flopped over the side, often threatening the street furniture. Some drivers were issued with poles to poke them back into place. Towards the end of the war—in January 1918—a Gas Restriction Order came into force making gas for motor vehicles subject to the same restrictions as petrol. A 'gas permit' was now necessary—and this had to be obtained from the Petrol Control Department of the Board of Trade. Southdown went back to petrol.

Alfred Cannon, meanwhile, saw little of the gas-bags. From 1915 to 1919 Douglas Mackenzie was both traffic and general manager whilst Cannon was a lieutenant in the Royal Engineers. Much of his time on war service was spent around Arras, planning layout and operating the rolling stock with the Railway Operating Division—his pupilage with the Great Western Railway having impressed the military rather more than his subsequent road experience. On the home front, Mackenzie kept things moving; a garage was built and brought into use in the autumn of 1916 at Freshfield Road and, following the influx of the previous year, an annual trickle of additional vehicles provided an acceptable level of service. He also convinced Tommy Sturt, the blacksmith, that he should work solely for Southdown. As all good blacksmiths do, he went home each day, summer or winter, in his shirtsleeves—a performance much admired, but not copied. For the first time, women were taken on as conductresses, and Bessie (Hilton) Winters, who signed on at Newhaven, was to remain with the Company until 1952, when she retired from the Express and Private Hire department at Eastbourne. The ladies were also employed as clerks and typists from 1915. Their rate of pay was two shillings and sixpence (12½p) per week when they started, rising by sixpence (2½p) per week to six shillings (30p)—rather less than their male counterparts now fighting the war and, although they enjoyed concessionary fares, this was at a time when the single fare between Brighton and Newhaven was one shilling (5p). Yet handwritten letters were now replaced by typewritten ones; the ladies were cheaper and more professional—and they've been there ever since.

In deference to marauding Zeppelins and Gothas street and vehicle lighting was now restricted and under the wear and tear of wartime traffic, together with monetary priorities deemed to be elsewhere, the condition of the roads—so recently improved as never before— began to deteriorate alarmingly. In Mackenzie's opinion, the worst "in the whole of Sussex ... was Hove Front, from the Halfway House at Portslade to Hove Street", with potholes over eight inches deep. The springs of the buses on the Brighton-Shoreham service were tested to the limit along this stretch. There was no through service between Brighton and Worthing at the time because an inspection of the chains of the Norfolk (suspension) Bridge over the estuary of the Adur at Shoreham proved them to be nearly rusted through. Brighton and Worthing-based vehicles ran as far as the bridge, where the passengers and conductors with parcels to exchange risked the gap on foot. Farther west, utilising an Act of 1878 aimed at recouping the damage caused by heavy steam-powered vehicles, Westhampnet Rural District Council tried to impose a 3d (1p) per bus mile charge on Southdown in October 1916 for each journey made on its newly-opened service between Bognor Pier and Portsmouth. Southdown promptly withdrew the service and appealed to the Local Government Board. Whilst Hampshire made no charge for use of the roads, West Sussex County Council followed Westhampnet's example and tried to charge Southdown and the Aldershot & District Traction Company the same amount when, in November 1916, they tried to open a service between Midhurst and Chichester. this had the effect of curtailing stage-carriage development in that part of West Sussex until after the Armistice.

The gap between Bognor and Worthing was not to be filled until the end of 1919, but the general line of demarcation for the northernmost spread of Southdown was settled as early as 1916, when a territorial agreement was signed with the East Surrey Traction Co Ltd (a forerunner of London Country Bus Services). Southdown's territory— as far as 'East Surrey' was concerned—was south of a line Haslemere-Slinfold-Horsham-Handcross -Crowborough (in other words mostly marginally south of the Surrey county border), with some free-range country between Uckfield and East Grinstead. Apart from some extensions into Surrey, of later date, that is where it remains. Connections with the East Surrey company's services dated from April 1916.

As the war came to a close in 1918, agreement on territorial boundaries had yet to be made with Maidstone & District Motor Services Ltd in the east and with what would become its neighbour in the west, Bournemouth & District Motor Services Ltd (Hants & Dorset Motor Services Ltd from 1920). These would present no problem: Walter Flexman French was chairman of both.

The Sussex coast enjoyed renewed popularity after the 1914-18 war, not only as a holiday and excursion centre, but, increasingly, as an area in which to live, and Southdown prospered accordingly. In this scene in Worthing No. 64, a Tilling-Stevens TS3 with Harrington 32-seat charabanc body placed in service in 1919 loads for the Long Furlong circular tour, with No. 14 (CD 5214) a Daimler Y-type bus with Dodson body of the same age behind and an unknown single-decker passing in the opposite direction. The charabanc had originally been registered IB 864 but within a few months had been re-registered CD 4864, as shown.

Chapter Four: Territorial development until World War II

The enforced division of Brighton and Worthing-based operations in WWI caused by the closing and replacement of the Norfolk Bridge at Shoreham, together with the isolation of the new depot at Bognor, doubtless had something to contribute towards the concept of district operational areas which now followed. Each of these would have a central garage and administrative facilities capable of coping with most local contingencies. They would be responsible for the day-to-day running of their own series of routes, which would fan out and inter-connect with services administered by the neighbouring depot.

In 1919, Alfred Cannon returned from his military service and both he and Mackenzie resumed their original duties within the company. Cannon's policy on local management was simple: wherever possible new posts were to be filled by those who had proved not only their ability but their loyalty to the earlier enterprise at Worthing and at Portsmouth. Douglas Mackenzie now set about his task as traffic manager in earnest; setting many examples which in less expansionist times have proven extremely difficult to follow. To some extent his efforts were aided by an unforeseen development: despite the dismantling of army camps, such as the particularly large one at Seaford, traffic actually increased. Numerous servicemen, having sampled the bracing and beautiful Sussex coastline for the first time, decided in the peace which followed to bring their families and settle there. Numerous 'bungalow towns'—some of them comprised almost entirely of retired railway-carriages— sprang up along the coast and began to reach out towards each other in classic 'ribbon-development' style. Lack of effective planning control resulted in some far from aesthetically pleasing additions to the landscape, but in the eye of beholders Mackenzie and Cannon this was beauty indeed. Within the first years of the 'twenties over 150 new vehicles had joined the Southdown fleet to help cope with the addi-

The Tilling-Stevens TS3 Petrol-Electric model was chosen as a basis for most of the double-deckers operated by Southdown in the period shortly after the 1914-18 war. Here No. 68 (CD 4868) is seen soon after entering service in July 1919 with No. 54, at the time registered as IB 854, a similar vehicle of which the chassis had been acquired in 1917, both having newly-built Dodson 36-seat bodywork. Number 54 had originally been owned by Thomas Tilling Ltd and registered LH 8926, becoming IB 614 at first with Southdown. It had thus already had three registration numbers but, like most of the other surviving IB-registered vehicles, received a CD number, in this case CD 4854, later in 1919.

Among the oddest-looking of Southdown's buses in the post-World War I era were the 'milk-churn' double-deckers, one of which is seen in the centre of this view at the Aquarium terminus then in use in Brighton. Number 118 (CD 5118) was a Leyland N-type placed in service in 1919 with a Dodson body which had originally been on the 1913 Daimler CC model, No. 7, transferred from Worthing Motor Services Ltd and shown on page 17. After a spell on a Tilling-Stevens TS3, No. 54 (seen on the previous page) in 1917-19, it had been rebuilt from 36-seat to 42-seat with the milk-churn compartment, wider than the remainder of the body, visible in this view. The Daimler single-decker nearer the camera is on the long service 31 to Portsmouth. The third bus is thought to have one of the ex-London & South Coast Haulage bodies.

tional call for transport.

Elsewhere, Mackenzie established routes which went inland, increasing the number of departures, even where at first there was insufficient traffic. As he had guessed, this created custom from sheer familiarity and the buses seldom ran empty. When they did, losses were more than recouped by profits made on established routes. Such repetitive bet-hedging built the framework and then infilled the Southdown network of stage carriage services; each route anchored in sizeable centres of population to ensure traffic in both directions. Mackenzie was able to claim "If there's someone walking along one of my bus routes, there's something wrong with the service".

Bognor and Chichester

On 13th May 1915, the new Company's purchase of Arthur Davies' licences and three Commer charabancs gave it a base in Bognor before the firm was properly registered. This was located in an office at Beach House on the sea front. Despite the local council's attempt to slap a mileage charge upon the Company, from this humble beginning there grew a distinct operational area with garages and outstations as far north as Midhurst and Petworth, where connections were made with the Aldershot & District Traction Co Ltd, and at Chichester, Compton, Eastergate, Selsey, Singleton and Wittering.

Arthur Davies had been in the motor car hire business since 1903. When Southdown made him its first Bognor manager it printed a special series of tickets for his motor trips which met their previous owner's ideas on how such things should be done. No conductors were employed on his Commers and the tickets were collected by the driver before starting on the return journey. Davies had used a similar system since 1908 and the early Southdown tickets actually named him as Bognor manager.

Meanwhile, it was decided that Bognor would act as the anchor-point for the stage-carriage service, first to Portsmouth and then eastwards to Worthing, together with another one which was to run to Worthing via Arundel. Chichester, which had enjoyed a double-decker bus service as early as 1906, when the Chichester & Selsey Motor Omnibus Co Ltd began its three-year life, had its service to Selsey restored by Southdown when George Edwards and two colleagues from Bognor started running one bus on a two-hourly headway from 7th July 1920. Twelve years later this service had grown sufficiently strong to run that quaint little train service 'The Hundred of Manhood & Selsey Railway' off its rails and into history. An office was opened at Chichester in June 1921. The route to West Wittering— originally motorised by George Theobold with a Siddeley-Deasy 14-seater on 5th November 1919 to the accompaniment of fireworks thrown beneath its wheels by supporters of the village horse-bus— was added following the acquisition of Alfred Trickey's business. Trickey, whose garage was at Birdham, owned a red Vulcan VSD saloon which entered the Southdown fleet in June 1923 and was quickly turned into a charabanc. Based thereafter at Portsmouth, Trickey was to serve with Southdown for over 40 years.

In November 1923, C. G. Shore of Bognor contributed his 'Royal Blue Services' Aldwick to Middleton stage carriage service, an excursion licence, three Guy B saloons and three Ford T 18-seaters to the Southdown operation; and two months later, at a time when the Company had only two buses based at Chichester, the purchase of W. G. Dowle's 'Summersdale Motor Services' between Chichester and East Dean

Acquisitions of independent operators helped in expanding Southdown's route coverage. Largest of the six vehicles from the Royal Blue Services fleet of C. G. Shore of Bognor taken over in November 1923 was this 1921 Guy B-type, BP 6985. A further extension of the classified fleet numbering system was introduced the same month, with some small Vulcans purchased new by Southdown given numbers from 301 upwards while the ex-Shore vehicles were given numbers beginning at 321 for the three Guys (this vehicle having that number visible on the frame behind the front mudguard) and from 331 for three Ford Model T. Note the informative side route board of characteristic Southdown layout.

Bognor's Station Road garage in the 'thirties. It was built in 1919 but had suffered a disastrous fire in 1923. The Tilling-Stevens Express B10A2 was Southdown's standard single-deck bus of the 1928-31 period, the vehicle shown, No. 660, dating from 1929 and having Short Bros bodywork. Its destination display is set for the Bognor town circular service.

added a Ford 1-ton model, two further Vulcan VSD saloons and a small office at the Dolphin Hotel in the city. A curiosity of that service had been the issue of ticket sets of a different colour for each of the three conductors. Dowle was to retire from Southdown in 1948 as the Company's manager at Chichester. It took the stringent regulations imposed by the Road Traffic Act 1930 to bring George Tate's 'Red Rover Motor Services' licence for the stage carriage service from Bognor to Pagham under the control of the Company in October 1932. Throughout this period—and until 1950—E. A. Cooksey was manager at Bognor, his thirty-odd years in that post constituting something of a record among local managers. Shore and Dowle's vehicles helped recoup the losses suffered when, in the early hours of 9th July 1923, the Bognor garage burned to the ground, destroying fourteen buses in the process. Vehicles brought in from other depots arrived in time to start the local services and not a journey was lost. Southdown had passed the first test of its abilities to cope in an emergency of such magnitude.

Early in September of 1934 Bognor Regis bus and coach station was opened on what had been the site of the old municipal offices and for its time, had well-appointed offices and public accommodation and a garage with 'no supports' in the modern manner. Chichester, however, was to wait another 22 years before it got similar facilities. Although Southdown had a garage in Northgate, and, from March 1926, an office in West Street, by 1938 'as many as six or eight buses at a time' hovered beside the cathedral to the consternation of local councillors. During the early years of the war the depleted ranks of cars because of fuel rationing nearly led to the establishment of a small bus station on the road-side West Street car parking area, but this was not to be.

G. Tate's Red Rover Motor Services of Bognor, with a stage service to Pagham was not taken over until October 1932. This solid-tyred 16-seater, BP 9579, dates from about a decade earlier; no vehicles were taken into Southdown stock.

Portsmouth

Waiting open-armed at Portsmouth for Mackenzie and Cannon's triumphant return was, of course, Frank Bartlett. Southdown's conductors paid in at his office near the Theatre Royal whence publicity, timetables and tickets were issued. Its vehicles were parked at various sites nearby until, in 1923, the Company opened its own premises in Hyde Park Road (now Winston Churchill Avenue). Before this depot opened, any vehicles arriving spare at Bognor from Worthing were sent on to Portsmouth on Sundays and Bank Holidays, where Barlett loaded them with 'day returns' to Bognor at five shillings (25p) per head. Five or six vehicles in addition to the normal hourly service would be despatched and one eye-witness claims to have counted 90 passengers off an open-topper at the other end of the journey. Waiting for the return journeys passengers stood good-naturedly in queues for up to three hours, and the last got home after midnight.

In the Portsmouth area particularly women enjoyed their newly-won emancipation and hauled their menfolk off into the surrounding countryside by Southdown bus. In the evenings the flow reversed, as increasing numbers of 'picture palaces' brought people into Portsmouth from all the neighbouring towns and villages, the cinemas quickly upstaging the old-established theatres. Despite economic troubles on the horizon, these were simple pleasures which thousands of local folk could afford and Southdown's resultant growth on Portsea Island and the surrounding area was rapid. Portsmouth Corporation Tramways Department was not quite sure what to do about this increased traffic into the town and at first asked for, got and enforced a 6d (2 ½p) minimum-fare ruling.

During May 1921 Southdown's neighbour to the west, Hants & Dorset Motor Services Ltd, had established itself on a route to Fareham, where it later opened an office at 12 West Street, and SMS Tilling-Stevens saloons and the occasional double-decker were despatched there to make connections with Hants & Dorset's Leyland saloons from Southampton. June 1922 saw this become a joint Portsmouth-Fareham-Botley-Southampton service (No. 35), terminating at the aptly-named Sussex Place in that growing commercial port—and subject to a 6d minimum fare there also to protect Southampton's trams. An occasional Saturday traveller on this service was a Southsea prep-schoolboy and his equally anonymous classmate who, according to his own much later account, (see his first autobiography, *The Moon's a Balloon*) visited Southampton to do some 'shopping' in that town without the benefit of money. He was to become the actor, David Niven.

A similar joint-service (No. 36) with Hants & Dorset started running to Winchester in December 1922 with driver A. M. Mountifield at the wheel of one of the first Southdown buses on the route. Portsmouth Corporation relaxed its minimum fare policy in 1925 and both companies were able to charge penny fares on five stages between Victoria Hall, Portsmouth, and First Avenue, Wymering. But then came the last of the boundary agreements which—with one later exception—sealed

The Southsea Tourist Co Ltd was Southdown's largest acquisition to date when taken over in February 1925, though the associated South Coast Tourist Co based at Littlehampton purchased just over a year earlier had already justified the beginning of a new 341-up series of fleet numbers for its seven Dennis vehicles, some of which came from the same batches as in the Southsea fleet, which took this series of Southdown numbers up to 376. Most were 3-ton models and one of those with bus bodies is seen shortly after the take-over, still in Southsea Tourist's grey livery. The Dennis chassis were sufficiently well regarded for several from both fleets to be rebodied in the 1926-28 period, some having their frames lengthened and later Dennis E-type engines fitted at the same time.

The newest vehicles in the Southsea Tourist fleet were two 51-seat open-top double-deckers on the longer Dennis 4-ton chassis placed in service in 1924, and registered TP 29 and 30. One is seen awaiting departure from Southsea on the route through Waterlooville to Horndean which competed with the Portsdown and Horndean Light Railway's electric trams, which became Southdown service 41, as shown above. These two buses remained in service with Southdown until 1930.

Southdown's western frontier for the best part of 60 years. In February 1926, both companies retreated to their original terminus at Fareham and passengers wishing to travel beyond that point changed to the other company's buses. A new series of 'through-tickets' was issued in celebration of this retrograde step, which handed back considerable traffic to the Southern Railway.

Believing that Southdown could be suitably sweetened by the removal of the 6d minimum-fare ruling, in the summer of 1925 Portsmouth Corporation began running its own buses beyond Cosham to Drayton and charged 1d (0.41p) for the journey—against 2d (0.82p) by Southdown. The management of the latter, not noted for its insouciance, retaliated with a penny fare over the same section and proceeded to saturate the route between the Theatre Royal and Cosham with vehicles in excess of available custom. Over some stages, the Company's buses actually charged less than the trams. Portsmouth Corporation then sought to impose conditions upon the continuing use of licences to what it called 'the private companies', and when Southdown appealed against this, the Ministry of Transport decided that the tramways would be suitably protected if Southdown charged one penny in excess of the tram fare. Whilst this ruling was awaited, April-December 1926, Southdown buses actually ran unlicensed within the Borough of Portsmouth.

Meanwhile, in 1925, Southdown had gained the Fratton Road route to Southsea and consolidated its portion in Portsmouth with the purchase of another company administered by Frank Plater, the Southsea Tourist Co Ltd (for the first, see 'Worthing' section this chapter).This was the largest firm that Southdown had taken over by that date and it was necessary for the Company to incur a bank overdraft to complete the deal. Transferred to Southdown on 1st March 1925 were 70 employees; a garage in Clarendon Road, Southsea; the tours office at North End; stands on the sea front; 30 Dennis vehicles; and the goodwill and licences for tours, express and stage-carriage work. For some while after the take-over, the ex-Tourist crews continued to wear their grey uniforms, and its personnel seemed well-pleased to be part of Southdown; 25 years later 27 of them would still be in the employ of the Company. Plater had run eight stage-carriage vehicles, including two double-deckers, on a route—with variations—to Waterlooville, Horndean and Petersfield. Over much of the route this service of pale grey buses had been in competition with the sea-green and cream tramcars of the Portsdown & Horndean Light Railway. Southdown's arrival doomed the trams by much stronger opposition.

Taking much of the weight from Frank Bartlett's shoulders throughout these days of expansion was Benjamin H. Ogburn who signed on at Hyde Park Road in July 1923 as cashier and assistant to the local manager. He had undertaken a rudimentary training in transport with Provincial Tramways' other Hampshire undertaking, Gosport & Fareham Tramways. He was also to take much of the detailed work off the shoulders of subsequent Portsmouth managers: Walter Budd, later general manager of Southern Vectis, J. B. Chevallier, A. F. R. 'Michael' Carling—a future general manager of Southdown—and R. Wakeling for the best part of 30 years, and was able to boast that he had "trained area managers very satisfactorily". During the latter part of the 'twenties Bartlett was thus able to concentrate upon being, in effect, traffic superintendent for the Portsmouth area of the Company. As such, he was able to organise further expansion and the provision of Dennis chasers against Millard's Safety Coaches who ran some GMC saloons from Fareham, West Street, to Porchester Castle and, from the Portsmouth end, against 'Portsmouth & District' who ran services as far afield as Denmead, Fareham, Hambledon and Westbourne. In both cases the operators simply withdrew from the route without need for Southdown to effect a purchase.

It seems to be part of the British character that requires an exceptional event to get people talking one to another. In September 1929, it was the international speed contest for seaplanes at Spithead, the Schneider Trophy, which provided the spur: for a week, both Corporation and Southdown buses beavered away harmoniously to shift the crowds of spectators who saw the Supermarine RAF team win the trophy outright for Britain. Again both operators strove for a working arrangement in the city and, on 8th October, a re-negotiated agreement was announced. Buses from both undertakings were to be run in accordance with timetables agreed by the respective managers. Tickets for issue on Southdown buses within the city were to be printed by the Corporation and all such fares were to

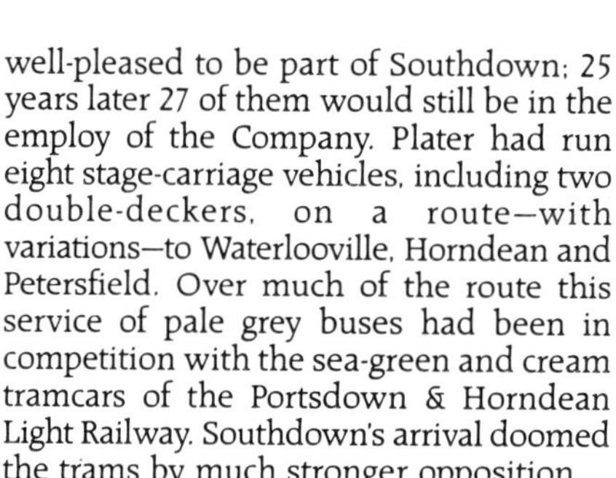

be paid to the Corporation transport department. The Corporation were to pay to the Company a sum per bus mile run by Southdown buses within the city, equal to the sum per bus mile shown to have been the all-in working expenses of the Company each year. Transfer and season tickets were to be continued and the fares charged on Southdown buses fixed by the tramways general manager.

The 'joint omnibus service' came into being on 1st January 1930. People living within the city thus gained a much more frequent and efficient service, but on outward journeys they tended to crowd out Southdown's long distance passengers. The 'cordial spirit of co-operation' lasted exactly two years: on 13th November 1931, Southdown appealed to the Traffic Commissioners against eight licence applications by the Corporation and had nine of theirs objected to by the municipality. A compromise solution pulled all the Corporation buses back south of Cosham railway gates—beyond which they had previously strayed to Drayton—and removed objections to Southdown working within the city on condition that protective fares should be reintroduced. These were to be on a sliding scale of 1d -2d (0.41p -0.82p) in excess of the Corporation fares, thus affording the municipality a degree of protection and—as it turned out—reducing the problem previously encountered by long-distance travellers. Despite a protracted argument in 1934, this arrangement remained in force throughout the 'thirties and until after World War II, by which time the trams which the Corporation had originally sought to protect had long since been replaced by the rapid and popular trolleybus.

Other 'private' operators within the borough still had to charge 6d (2½p) excess. Not included in their number was one which in the spring of 1929 had set out to challenge the growing weight of Southdown and lost the battle in the space of twelve months. The story is of particular interest, because it took the co-operation of a sister company to achieve victory, and the final outcome was not quite what its ally had expected, since in 1924 Southdown had already given up the territory in its favour after just one year's service.

C. R. Fuger of Warsash, on the River Hamble and well within what was to become the area of Hants & Dorset Motor Services Ltd, started running buses to Fareham from his home village in 1919. Ten years later he'd come to the conclusion that the majority of his passengers went on to Portsmouth in Southdown buses and, to the Company's disbelief, applied for and was granted licences to run a total of 28 journeys per day at half-hourly intervals (against 44 by Southdown) beyond Fareham to the Victoria Hall, Portsmouth—which he now set out to do. Fuger's 'Warsash & Fareham District Bus Service' as it was called was ill-equipped to deal with the counter-punch which now followed. Hants & Dorset withdrew from the route; Southdown placed two brand-new Leyland Titan double-deckers in an old strawberry-packing station which it acquired at Fleet End, near Warsash and proceeded to saturate the entire route from Warsash to Portsmouth with 32 return journeys per day and with the fare on the section to Fareham down to 6d (2½p) return. Fuger capitulated on 1st May 1930, selling his fleet to Hants & Dorset for considerably less than that company had originally offered. He transferred the title of his Warsash to London express coach service to his wife, Mrs M. V. Fuger, but this, too, was to be sold in 1935—to Southdown (see Chapter 7). Southdown, however, remained on the Warsash route well within Hants & Dorset territory, with whom it now ran the service jointly west of Fareham. It was to remain on the Warsash road to this day.

Another acquisition in 1929 was made far less traumatically and involved the transfer to the Company of the licences and four Dennis vehicles of W. E. Pinhorn's 'Meon Valley Services' of Catherington, which operated a stage-carriage service into Portsmouth from East Meon and Clanfield. Southdown settled for the Clanfield to Portsmouth section. Three years earlier it had consolidated a place upon Hayling Island by purchasing Holt's Motors' Rowlands Castle to Hayling service; together with a stage-carriage service from Hayling to Waterlooville and a small Morris saloon bus from W. Stride. On Monday 1st November 1926, Frank Bartlett had launched its Hayling presence in some style. He took the councillors of Waterlooville on the inaugural run and entertained them to tea at the Ship and Bell Hotel, Horndean. Hayling had enjoyed a railway service since 1867, but despite vigorous efforts to develop the island, including the establishment of a race course, it remained something of a retreat, the weight of traffic considerably curtailed by a road bridge which later imposed curious operating conditions upon Southdown. Despite the building of a grand Victorian parade on the southern shore, all that followed was an extremely modest series of seaside amusements. Two other small operators—P. C. Belier and G. W. Meekings—had, however, found enough seasonal traffic to offer stage-carriage services along the sea front, and the licences for these were acquired by Southdown in July 1933. Well used to their own sea front, a sufficient number of Portsmouth and Southsea's citizens now proved interested enough in what Hayling had to offer to make an island-to-island service a viable proposition.

Most fascinating of the contests for custom, however, was that going on between Southdown and the tramcars of the Hampshire Light Railway (Electric) Co Ltd. The trams, which left Cosham on an embankment and a series of bridges over both railway and roads, climbed Portsdown Hill at a speed which had the edge over the rival buses, ran along roadside track in the countryside and up the middle of the road through the townships to Horndean. They had their loyal supporters, but it was the ability of the buses to make route variations and go beyond the tramway termini which eventually won the day for Southdown. The tramway was purchased by the Company and Southdown buses took over the route on 10th January 1935, leaving the parent 'Provincial' company with only its Gosport & Fareham operation to represent what had once been the major provider of road transport on both sides of Portsmouth Harbour.

Walter Budd having gone over to the Isle

The Leyland Titan double-decker sharpened Southdown's competitive edge considerably. As well as being the instrument by which the enterprising C. R. Fuger was defeated, as described above, its lively performance enabled more traffic to be taken from the Portsdown and Horndean Light Railway operated by the Hampshire Light Railway (Electric) Co Ltd. This scene, with two 1931 Short Bros-bodied TD1 models, Nos. 927 and 930, was specifically posed in 1935, prior to the withdrawal of the tram service. Both Titans were among those sold to Wilts & Dorset in 1939, serving that company longer than Southdown.

Fewer vehicles were taken into stock from the businesses acquired in the 'thirties but eleven Thornycroft buses came into the fleet from F. G. Tanner's Denmead Queen concern of Hambledon in March 1935. This one, a UB model with 26-seat body by Wadham Bros, was the oldest, dating from 1927 (when the picture was taken). By this period two-digit numbers were generally used for 'odd' vehicles that did not fit into recognised series and the ex-Tanner fleet was given this series, from 39 (this vehicle) to 49, though 500 was added to the numbers of the survivors in 1936 to make way for Leyland Cub models.

of Wight to take up the general managership of the Southern Vectis Omnibus Co Ltd, supervision of activities in Portsmouth was now the responsibility of J. B. Chevallier. To him fell the task of securing for Southdown a complete monopoly of stage-carriage traffic from Portsmouth to the surrounding countryside. In March 1935, F. G. Tanner of Hambledon at last agreed to part with his licence for the World's End-Hambledon-Cosham-Portsmouth route of his red-liveried 'Denmead Queen Motor Services'. The eleven Thornycroft saloons which entered the Southdown fleet represented the largest single purchase of another operator's stage-carriage vehicles in the Portsmouth area and firmly established the Company in another village with historical connections for Mackenzie and Cannon. A clean sweep of the area followed in September with the acquisition of the licence for the Meonstoke-Droxford-Wickham-Southwick (Hants, pronounced 'Suthick')-Portsmouth & Southsea Service of Blue Motor Services (Southwick) Ltd of Boarhunt, to the north-west of Portsdown Hill. This all helped to change Portsmouth Corporation's views on co-operation with Southdown and it now began to think seriously about a proper co-ordination agreement, particularly since the Company's operations had grown so large that, in 1934, an impressive new garage-cum-coach station had been opened at Hilsea on the northern edge of Portsea Island. But, as Southdown's businesslike fleet of Leyland Titan double-deckers got steadily larger, the Corporation concentrated upon the replacement of its trams with trolley and diesel-engined buses. The territorial ambitions of Nazi Germany were to postpone such an arrangement until an even later date.

When war was once again declared in September 1939 there was little doubt that Britain's major naval base would attract the attention of the Luftwaffe. There was immediate work to be done in the evacuation of children from the city to the surrounding countryside—although during the 'phoney war' many of them drifted back again. After Dunkirk, Portsmouth was to share with Southampton in excess of 1,500 air-raid alerts with extensive bomb and land-mine damage both in and outside the Dockyard. Vehicle-dispersal at night removed buses from the menace of incendiary bombs, but during the particularly severe raid of 10-11th January 1941 two of the Company's coaches which had been converted into ambulances were destroyed. The following year saw the re-engagement of conductresses, and one, Madge Joslin, was to be appointed inspector some 33 years later.

Eastbourne

Eastbourne is, of course, somewhat famous in the transport world as the municipality which was the first to gain parliamentary powers to run a motorbus service. This it commenced to do on 12th April 1903, after representatives of the town's Motor Omnibus Sub-Committee had gone over to Hastings to see a Milnes-Daimler at work and had been much impressed by it. Recent research suggests this vehicle was among those which Douglas Mackenzie later purchased for use at Worthing. Eastbourne's buses ousted the horse-buses of F. Bradford and William Chapman & Sons, the latter turning to excursions and tours with wagonettes and brakes. Chapman had thus captured the lucrative Beachy Head traffic, despite the efforts of Mackenzie, on behalf of the Sussex Motor Road Car Co Ltd, who in March 1908 had unsuccessfully approached the Corporation for licences to run motor coaches to the headland, some 575 feet above sea level. Chapmans had also secured for themselves what would prove to be a quarter of a century's reign as the town's leading proprietor of excursions and, latterly continental tours (see Chapter 6).

As early as July 1915, Southdown had run motor coach excursions and a short-lived twice-a-day bus service to Eastbourne. It arrived in the town as fully committed stage-carriage operator in 1920, when pockets of post-World War I settlement along the coast developed into the positive housing boom which was to make the Company's march eastward from Brighton a major feature of its rapid expansion. A completely new community with a name now evocative of those heady post-war years sprang forth, and sought transport.

It was called Peacehaven—and its brave-new-world citizens wanted to go into Eastbourne, as well as Brighton. Southdown now provided the means by extending its Brighton to Seaford route. Once in Eastbourne, it started a service which went north west to Hailsham and Uckfield and another which strode on eastward to Bexhill and Hastings, run jointly with Maidstone & District Motor Services Ltd. Additionally, for some two years, Southdown buses turned off at Ninfield and went to Battle—an evocation of somewhat earlier vintage. These services passed along the Polegate to Hailsham road, considered at the time to be the worst in the country. The county councils faced a serious dilemma: WWI traffic, steam traction engines and a lack of money to spend had ruined many roads, which were now heavily pot-holed. To run the rain off those that they repaired, the roadways departments gave them a heavy camber. This meant that double-decker buses in particular leaned heavily to the side, where they knocked street lamps out of their holders and swept away awnings from shopfronts. To counteract this drivers kept to the middle and got accused by irate fellow motorists of hogging the road. Solid tyres wore tracks in the flimsy road surface and in places it was necessary for the bus to be put into a lower gear in order to make headway **downhill**, or to turn to left or right. Only the combination of a good tarmacadam crust and the introduction of pneumatic tyres for the buses from the middle 'twenties onwards solved a problem which posed difficulties for everyone concerned. East Sussex lagged somewhat behind West Sussex in its efforts to repair its roads, despite the latter's unsuccessful attempt to impose a mileage charge upon Southdown. At one stage, in 1921, it actually suggested closing the coastal road between Seaford and Eastbourne to anything heavier than a private motor car. Southdown, however, continued manfully to strain its springs along these routes until East Sussex council caught up with the rest.

As befitted its foundation upon the estate of the Dukes of Devonshire, Eastbourne was

The pride of Eastbourne garage in 1934 were the two new six-wheel Leyland Tiger TS6T models for the popular service to the top of Beachy Head, won in an early traffic court hearing in 1931, No. 50 (later 550) being seen when new. The aim was to maximise the seating capacity whilst not infringing the restriction to single-deckers imposed as a condition of the road service licence. Accordingly the Short Bros bodywork seated 40 passengers but the appearance was very similar to the latest express-service coaches and, as with them, a folding roof was fitted. Originally petrol-engined, they and two similar vehicles on the almost identical TS7T chassis were converted to oil (diesel) in 1940. They remained in service, latterly with seating reduced by one, until 1952, by which date open-top double-deckers were permitted.

found to be attracting a rather more staid holidaymaker, eager to go farther afield with a purpose, usually to see something historical -and to see it in comfort. Accordingly, Southdown's excursion traffic from the resort grew apace, and the majority of the numerous acquisitions of rival businesses was in the excursion and tour sector. The exceptions were T. A. Piper's 'Red Saloon Motor Services' stage-carriage service into Eastbourne from Hellingly, and H. J. Twine's Hailsham-based fleet of four saloon motorbuses together with services from Eastbourne to Jevington and to Polegate. Both operators gave up their licences in favour of Southdown in 1929, the year that it purchased the Old Lion Brewery site to build the Pevensey Road bus station. Post-Traffic Act difficulties caused J. Haffenden of Vines Cross to follow suit with his Heathfield to Eastbourne service in March 1932.

Despite a mildly unhappy start at Eastbourne, when the municipality sought unsuccessfully to set and regulate the times of departure for the Company's buses running out of the borough, Southdown was to establish an extremely cordial relationship with the town. However, whilst some routes were clearly of no interest to Eastbourne Corporation's Motor Omnibus Department, such as those which the Company was to establish to Heathfield and the extremely long run to East Grinstead, others like the Stone Cross circular, the Wannock and Jevington, and the Pevensey Bay routes had their termini tantalisingly close to the municipal boundary. The most contentious route proved to be that to the top of Beachy Head, so long the territory of Chapmans. In August 1931, the Traffic Commissioners for the South Eastern area sat at Eastbourne and heard applications for stage-carriage licences for this route from six operators including the incumbent Chapman & Sons, Eastbourne Corporation and Southdown. The latter proved to be the successful company, on the condition that they used only single-decker buses (later relaxed to include open-topped double-deckers) and did not take up and set down the same passengers between the Redoubt and the foot of Beachy Head. In 1934 Southdown realised that the 'single-deck' restriction could be lessened by purchasing vehicles of greater carrying-capacity for this service, and invested in the first of four petrol-engined Leyland Tiger six-wheelers built to a suitably coach-like specification. By their presence in Eastbourne, embellished with their Mackenzie-style fleetname, they added to the Southdown image of 'something special'—and did good business accordingly.

World War II was to hit Eastbourne particularly badly. Declared a restricted area in 1940, the resort was an easy target for Luftwaffe aircraft coming in low from the Channel, and numerous bombs were lobbed into the town often without warning. With its holidaymakers temporarily lost and large numbers of its inhabitants evacuated, the municipality was to loan to Southdown ten petrol-engined Leyland Titan double-deckers of various marks and vintage. For some three years, these, together with others borrowed from East Kent, were to provide a welcome relief at Worthing and elsewhere to the difficulties encountered by Southdown following the requisition or converting into wartime guise of many of its vehicles. Despite the danger on the coast, it was inland at Heathfield that a bomb dropped directly beside a Southdown bus and blew it over a hedge into a field, killing five passengers and the conductor—the Company's worst incident of the war.

Worthing

It was to be from the premises at 23 Marine Parade, Worthing (see page 16), that Mackenzie and Cannon were to plan the rapid expansion of stage-carriage services into the Sussex and east Hampshire countryside. It was from here too, incidentally, that for many years, and together with their long-time friend Percy Lephard, they were to run the affairs of Wilts & Dorset Motor Services which they had taken over and registered as a limited liablility company on 4th January 1915—four months before the official launch of Southdown. The specifications for Wilts & Dorset vehicles, the arrangement of their colour schemes (albeit basically red) and the Mackenzie-style fleetname on its coaches was to last almost until World War II and in some cases into the post-war years. During that period, the Wilts & Dorset vehicles were sent to Worthing for overhaul and new ones were marked down to that address by manufacturers. The association was also to lead to the establishment of the South Coast Express coach service an excuse to use the name previously denied them.

Number 23 was a rambling premises: one was greeted at the entrance by notice boards listing the services and the times of departure. In the basement was Douglas Mackenzie's pride and joy— his ticket store. Arranged on its shelves in sequence and by route and price, each bundle was tied with string. Each piece of string was saved as it came off and was measured against a rule on the front of his desk drawer marked to represent the size of bundle it would hold—and was put into a compartment for further use. He wasted nothing. In the front room upstairs the conductors paid in their money and at the back the ladies typed and kept the waybill book. Mackenzie's office ran across the front of the building on the next floor above. In the attic, Philip, the regulator, and his wife lived on the premises, devoting their entire time to the Company. Acting as Worthing manager initially was chief cashier Parker. Upon his death in 1922, Inspector E. James was to begin his twenty-

When George Town of Worthing sold his old-established business to Southdown a week after Armistice Day, 1918, the rolling-stock comprised two of the X-type buses built in 1909-10 by the London General Omnibus Company before it began production of the famous B-type. These had proved under-powered, being capable of only about 14 mph, and withdrawal from the LGOC fleet began in 1914. The vehicle shown, LN 9967, had been X59 and became Southdown No. 90. When taken over, both had 36-seat Dodson bodies recorded as having originally been built for the Metropolitan Steam Omnibus Co—similar remarks apply to the Dodson double-deck bodies on Worthing Motor Services buses of the 1912-14 period. When the ex-Town X-type chassis were withdrawn, the bodies were transferred to other chassis, at first Tilling-Stevens and then Leyland N-type with milk churn compartment, like that shown on page 24.

year duty as local manager. Also working in the building was the man who would succeed him in World War II, Inspector W. J. Cooper, and Mackenzie's secretary, Miss Hilda Botting. The original garage, now used as the coach station, was located behind the Clarendon Hotel. Two further garages were constructed nearby as the business expanded.

Until Douglas Mackenzie married Eva, comparatively late in life, he always gave the Marine Parade office as his private address. Miss Winnie Bravant, who worked for Southdown from 1920 -1964, is of the opinion that he lived nearby in small hotels and spent more time in the office than he did anywhere else. He was always there first thing in the morning and the staff invariably left him there at night: he seemed never to leave the office for meals and appeared to live on bars of chocolate. He would take such a bar from his pocket and present it to a driver asked unexpectedly to do some overtime, "in order to keep him going". His emergence from the office was now restricted to crossing the road to listen to the engine of a bus the driver was doubtful about—an engineer to the last. Marriage changed his lifestyle somewhat. He found a new interest in his Scottish origins, joined Caledonian societies and went so far as to call his first proper home in Mill Road, Worthing, 'Scatwell'—a seat on the River Conon in Clan Mackenzie country.

James Town had died at the ripe old age of 86 on 1st January 1912. His son George kept the services running and did what his father could never bring himself to do—he turned to motorbuses. In June 1916, he placed his first on the Heene to Broadwater route, with his daughter acting as conductress. Fittingly, perhaps, he was the first local operator to run his buses on town-gas during the WWI petrol-shortage, and whereas most other operators carried the gas-bag on the roof, at least one of his was towed behind on a trailer. George Town finally brought to an end his family's long association with transport in Worthing by selling his three remaining double-decked open-toppers to Southdown on 18th November 1918. They had been operating on the Broadwater to Elm Grove route.

Having gained what at that time constituted a virtual monopoly in Worthing, the Company now set out to establish, with variations, a basic eleven routes in, across or starting from the resort with, of course, the famous route 31 going through all the way from Brighton to Portsmouth. On the morning of its inauguration Mackenzie had emerged from the office to wait, watch in hand, for the first arrival from each direction. He was not disappointed, remembered William Jay, conductor of the first Brighton-bound vehicle. Starting with Horsham in 1919, at the end of the route via Ashington, Southdown's Worthing office was to assume responsibilty for garages in eight other locations, ranging from the dormy-shed at Dial Post to what would eventually become a 55-car depot at Horsham.

Littlehampton depot had its origins in the office opened at the Dolphin Hotel in Surrey Street and supervised by Inspector R. Howe. From its initial two buses, Littlehampton was to gain a garage capable of housing 30 vehicles, in East Street, with H. E. Humphrey as garage superintendent. In 1924, Southdown took over the Littlehampton-Arundel-Angmering service of the South Coast Tourist Company Ltd— the first of the 'Tourist' companies controlled, in this instance with Norris Brothers, by Frank Plater—together with its excursion licences and eight vehicles. In its running battle with Southdown, Tourist had brought the fare

When taken over in August 1938, Tramocars Ltd of Worthing was operating a fleet of eleven S&D Freighter models of the type long familiar as a basis for refuse collectors, with tiller steering, T9 (PO 7706) dating from 1933 and having Harrington 26-seat bodywork. Two more vehicles of the same make but of a new rear-engined design had been delivered direct to Southdown the previous month, the entrance position in this case being central rather than at the rear.

Even if Southdown was largely run from Worthing in the early days, its official address was always in Brighton. This photograph of No. 94 (CD 6894), one of ten Tilling-Stevens TS3A models with Tilling bodywork delivered in 1922, marked the beginning of the use of a larger type of double-decker, with a total passenger capacity of 51. The design incorporated an enclosed windscreen and forward-facing seats for almost all passengers, basically similar vehicles being taken into stock until 1926. Despite the 'Littlehampton, Worthing and Brighton' board, the absence of a side route board suggests that it was a posed picture, with the party of schoolgirls as not unwilling models. All of the ten vehicles had been withdrawn by 1930 despite conversion to pneumatic tyres around 1927-28, though this one was sold for further service with the Tyneside Tramways and Tramroads Co via Tilling-Stevens.

from Littlehampton to Arundel down from 10d (4p) to 4d (1½p), at which figure Southdown retained it until 1951.

Back in Worthing that year of 1924, however, a new and what proved to be a minor trend-setting operator gained a stage-carriage licence to operate along the Front. Walter Gates, a Londoner, had returned from a lengthy visit to New Zealand and noticed that the elderly at Worthing had difficulty in mounting the steps on Southdown's high-framed buses. Accordingly he asked Shelvoke & Drewry, the manufacturers of the 'Freighter' refuse-wagon, to provide a chassis for bus work and from a garage in Wordsworth Road put the first of his curious little vehicles, which initially he drove himself, on service along Worthing Front. Some fifteen of these 'Tramocars', red with white roofs and gold lettering, were placed in service between then and 1935. Eleven of them were to survive long enough to be purchased by Southdown in August 1938. The latter continued to run them in decreasing numbers until 1942, when they were replaced by more modern conventional buses which by now had relatively lower steps and floor level. Older residents continued to call the replacement Southdown vehicles 'Tramocars', much to the puzzlement of visitors.

Meanwhile, Southdown's operations in the Horsham area increased following a spectacular but disastrous fire at the Barns Green garage of W. H. Rayner & Sons. Because of a curious allotment of their responsibilities, the Horsham brigade did not respond to the call and the Steyning brigade had to go fifteen miles to find aluminium and burning oil flowing down the road and a bucket-chain doing its best to save Rayner's house and furniture. Only one bus was saved from the seven-strong fleet and this, plus the goodwill of services from Horsham to Coneyhurst, Horsham to Brooks Green and the Horsham town service, Rayner sold to the Company in January 1935. That same month, T. W. Carter of Horsham relinquished his service to Steyning. The newly acquired routes were added to those gained in January 1933, when S. S. T. Overington, trading as the Blue Bus Service, sold his licences for the Balcombe and Maplehurst routes to Southdown. One further acquisition in the years before World War II was that of R. J. Smart's 'Ferring Omnibus Services' to Goring.

Brighton

Even if Cannon and, particularly, Mackenzie were reluctant to move eastward along the coast from Worthing, the official registered offices of the Company were, from the first, in Brighton. To start with they were at the 73 Middle Street premises once owned by 'Jolly Jumbo'. Some administrative work seems to have been undertaken also at 6 Pavilion Buildings, before part of Stuart Smith's house at 5 Steine Street was acquired for the role of head office. Whilst Frank Smith from BH&PUOC assumed the role of company secretary, Stuart Smith, previously manager at Brighton for WMS, took on the role of supervisor of the Company's clerical and private-hire work.

Initially traffic operations and the engagement of staff were undertaken by French's colleague, H.E. Hickmott, but when he went first to war and then north to Ribble, at

Preston, as Major Hickmott, his duties were taken up by Frederick Mantell—until his retirement in December 1932. In 1918, Stuart Smith moved across to the opposite side of Steine Street and made the whole of his house available to Southdown. The first workshops in Brighton were located in Upper St. James's Street, but the new (in 1916) Freshfield Road garage incorporated improved facilities. Any chance of competition which could have arisen as a result of the purchase of BH&PUOC's town services by Thomas Tilling Ltd was averted in February 1917, when Thomas Wolsey, representing the Tilling organisation, was elected to the Southdown board. He was the younger brother of Walter Wolsey, Thomas Tilling's son-in-law. At the same time Southdown was converted from a private to a public company.

In 1921 a joint service with Maidstone & District was started between Brighton and Hawkhurst, whilst other services had already been established from Brighton to Worthing, Eastbourne, Lewes, Lindfield, Horsham, Uckfield, Petworth and, of course, the long journey all the way to Portsmouth. A depot was opened in Haywards Heath in 1925, although operation had already been established from a dormy-shed at nearby Scaynes Hill. By the summer of 1921, it was possible to travel all the way from Brighton to London by stage-carriage bus: Southdown to Handcross - 2/6 (12½p); East Surrey to Reigate - 2/5 (12p); London General - 2/- (10p): a total fare of 6/11 (34½p). Although this took some 5½ hours it was considerably cheaper than the contemporary charabanc fare of £1.

The following summer, there began a controversy over Brighton Council's decision to grant tentative permission for Southdown to build a garage for '200 coaches' in the eastern end of the Aquarium on a 60-year lease. The town divided into two-hostile camps: "attend a meeting at the Dome and protest" ran the posters. Said one councillor prophetically: "Messrs Tillings are a very large shareholder in the Southdown and some day they may be a very large combine; not only the Southdown and Tillings, but the London General Omnibus (also). What would happen if we were to have thousands ... of people put down on the Brighton Front all day?" The Council decided by one vote not to proceed with the lease. So ended Southdown's first attempt to build a bus station—and in the long run the Company was never to embrace the notion of the need for bus stations as avidly as most of its sister companies.

The frustrated architect of the Aquarium garage plan was C. E. Clayton, who had designed Southdown's Freshfield Road garage and Tilling's one in Conway Street. He was one of those rounded men who was also an educationist, JP, reformer and chairman of Brighton College of Art. Another in similar mould was John Henson Infield, whose deep knowledge of the area was to lend renewed local strength to Southdown when he joined its board of directors in May 1924. Latterly, and until his death in June 1942, he was to become vice-chairman of the Company which enjoyed his talents legal, social and practical: sagacious and kindly, he was a barrister at law, JP, director of Brighton's 'Grand Hotel' and of the 'Theatre Royal' Brighton and chairman and managing director of the Southern Publishing Company. Interestingly, that firm had from the beginning printed and published Southdown's timetables and publicity. Meanwhile, in 1925, Southdown had moved into a further new garage built in Edward Street.

The General Strike of May 1926 had comparatively little effect upon Southdown. Whilst the Brighton trams and Tilling's buses were at a stand-still, Southdown had buses out along the coast roads and on country routes, but on the first day none ran out of Brighton until 2pm when they were driven by the large proportion of non-union men. They were haranged to some extent by pickets at the Aquarium pick-up point, and were actually guarded by police, but this was largely unnecessary for none of the strikers attempted to interfere with passengers.

But Southdown took a strong line with the strikers: "All employees who have not remained at work during the present crisis are regarded as no longer in the service of the Company. The insurance cards of these employees will be returned at once and they will be called upon to return their uniforms and other property of the Company". It announced that an agreement entered into with the crews' trade union on 27th April, now violated, was forthwith terminated and that in future it would not recognise any trade union or engage union labour. By 14th May, almost all the strikers had taken the opportunity to rejoin the Company as 'new employees' and the service had returned to normal. This had been standard BET policy: if the TUC had been ill-prepared for the strike, it met scarcely more effective opposition to trade unionism than that put up by BET's subsidiaries—and trade unionism didn't get a foothold in Southdown worth the name until after World War II.

Sidney Garcke, son of BET's founder Emile Garcke, had joined the Southdown board in March 1924 and had been elected chairman of the Company in January 1926, following the death the previous month of Walter Flexman French. In view of events taking place in the bus world at national level this would stand Southdown in good stead. In 1922, Thomas Tilling Ltd had taken a large holding in BET's subsidiary, the British Automobile Traction Co Ltd. This somewhat untidy marriage was smartened up in 1928, when Tilling & British Automobile Traction Ltd was formed—under the chairmanship of Sidney Garcke. This practically ensured that in any parting of the ways, Southdown and its chairman would end up in the same camp.

In November 1928 the four main-line railway companies started talking with Tilling & BAT about investing in its subsidiary bus companies rather than in their own road passenger services for which recent powers had been granted them by Acts of Parliament. Both parties approved the railway acquistion of shares equal to those held by Tilling & BAT in the appropriate bus companies. Thus the Southern Railway Company paid £3.5s.0d (£3.25) per £1 share for one third of Southdown's stock which stood at £200,000. It was considerably more than Southern paid for a similar share in East Kent and Maidstone & District. As a

Purchase of second-hand vehicles, as opposed to those taken over as part of a complete business, were very rare after the special circumstances of the 1914-18 war. In April and June 1927, however, a batch of twelve ex-Birmingham Corporation Daimler Y-type double-deckers were acquired via the Associated Daimler Co Ltd, the joint sales company set up by AEC and Daimler in a short-lived union—manufacturers frequently acted as dealers at the time. The vehicles had been quite extensively rebuilt, though the chassis originally dated from 1916. They had Brush bodywork built in 1922, and had later received windscreens and cab half-doors. They had also received AEC engines in 1925-27, Birmingham being an AEC stronghold at the time, and some also received AEC radiators, as shown by OB 2103, numbered 317 by Southdown and seen in Brighton on the North Moulsecoomb service in November 1927. Note the conversion to pneumatic tyres on the front axle only. They were kept for two years during a period when purchase of new double-deckers virtually ceased.

The appointment of Sidney Garcke, son of the BET founder, as Chairman in January 1926 on the death of Walter Flexman French, followed by the restructuring of Tilling and British Automobile Traction interests in 1928, might have been expected to signal some change in vehicle policy. The ordering of 23 Leyland Titans for delivery in 1929 after a couple of years when no vehicles of this make were added to the fleet, and the choice of Brush, with its close historical association with BET, as builders of the 51-seat open-top bodies might have been the beginning of a new era. Leylands were back to stay, but no further orders went to Brush. Note the folding door at the rear of the lower saloon in this official view of No. 804.

result of the agreement, the assistant general manager of the Southern, Col G. S. Szlumper joined the board in February 1930 followed by Ralph Davidson in September—the advance guard of numerous railway representatives until the establishment of the National Bus Company in 1972 rendered the system obsolete.

Incorporated in the Road Traffic Act 1930, was a group of clauses which brought thorough-going changes to the omnibus industry. These required the crews to be licensed centrally, the vehicles to be submitted to strict mechanical tests, each bus or coach service, tour or excursion to be separately licensed, with opportunity for objections to be aired in special 'traffic courts' by other operators or the railways, should they feel that their own trading prospects were at stake. The country was divided into 'traffic areas' administered by 'Traffic Commissioners', transgression of whose rulings would carry severe penalties. Independent operators saw such close control of their affairs as the beginning of the end, and a considerable number chose, or were obliged, to sell up soon after the Act came into being. Southdown, well represented in the courts by extremely sound legal advocacy came out well on balance and, in the nine years between the enactment and the outbreak of World War II, absorbed the goodwill of no fewer than 44 rival operators throughout its area; against 26 in the fifteen previous years. Apart from H. J. Sargent's East Grinstead-Ashdown Forest stage-carriage service, acquired in September 1937, such take-overs in the Brighton depot's area consisted of excursion, tours and express services.

If truth be told, the new legislation probably put an end to the vestiges of the more unusual activities engaged in by Southdown in its first fifteen years. In the early days, no reasonable offer was refused. Charabancs had been changed into lorries, by removal of their seats, and carried such diverse cargoes as shop-fronts, flour from mill to bakery, groceries to wholesalers, and even a totally unmanageable load of loose sprats for Billingsgate fish market. Two ex-RAF type Leyland lorries, which were later transformed into saloon buses, worked on the removal of chalk from the foundations of Brighton's 'Regent Cinema'. Stage-carriage vehicles on service had carried churns of milk into town from outlying farms, mail for the Post Office, pigs, goats or calves in 'pig-nets' and sides of bacon and carcasses of pig or sheep up the ladder and on to the roof-rack. Conducting such a bus was not just a question of collecting the money and issuing tickets.

Until the establishment of the London Passenger Transport Board in 1933, Uckfield was the meeting point of services operated by Southdown, the East Surrey Traction Co Ltd and Autocar Services Ltd. Southdown now gained some extra territory: services previously provided by East Surrey and Autocar south of Crawley and East Grinstead, not being part of the area prescribed for LPTB, were allocated to Southdown, and the co-axial point moved eastward to Heathfield, which was now shared with Maidstone & District.

Back in Brighton, the Steine Street coach station had been enlarged and Pool Valley was in use as the Brighton stage-carriage terminal. It had soon become clear that attempting to cope with the maintenance of the fleet at the comparatively new Freshfield garage was an overwhelming problem. Further land had thus been purchased at Portslade and a new central overhaul workshop had been opened there in 1928. From the beginning its main functions were dealt with in the chassis shop, the body shop and the paint shop—each with its own sub-sections and stores located around it. The integral office block became headquarters for the chief engineer—R. G. Porte, initially—the assistant engineer and the stores superintendent, who rivalled Mackenzie in his meticulous system for numbering the parts in their separate bins. The tailor's shop was also located at Portslade—the summer staff handing over their uniforms to a full-time tailor who cleaned, renovated and placed them in store for re-issue the next summer.

The 'seasonal' nature of Southdown's traffic also dictated the way the central workshops would operate. For instance, at the beginning of April 1938, 467 vehicles were licensed; in August 699, covering two-thirds as much again as the winter mileage of 1,500,000. During the summer every vehicle in the fleet, save those for sale, had to be on the road. In winter, the canvas-roofed coaches, which only covered some 10,000 miles each season, were either laid up or formed a spare-time docking job in the winter. They, and the Company's 23 open-topped Titans were de-licensed during the winter months, the comparatively light use of the latter enabling them to put in over 20 years service with Southdown. Fifty dual-purpose employees were one of the most important factors of the seasonal organisation. During the winter they worked on overhauls at Portslade; at the beginning of summer they transferred to the depots, to act as drivers, conductors or tradesmen. Other seasonal crews were given, on a rotational basis, the chance of permanent employment as vacancies occurred—and, being Southdown where long-service was the norm—this was not quite so often as some would have liked.

At overhaul time, the bodies were also refurbished, from uphostered seats outwards, but the days of Mackenzie's body-swaps had gone. Now they were never separated from their original chassis at any stage of their Southdown lives. Service buses exclusively were attended to in the summer, and work on excursion and tours vehicles took place in the winter. The works and facilities were, of course, added to over the years as further vehicles and new technologies demanded. Both the facilities and the level of craftsmanship of the workforce were second to none in the industry and, quite simply, that is the way it has remained. A 'standard-image' was never likely to go down well at Southdown;

The replacement of the Tilling-Stevens single-deck bus fleet took place in two main stages, in 1935-36 and more especially in 1938-39, when a total of 86 Leyland Tigers with Harrington bodywork were placed in service with fleet numbers beginning at 1400. This example, No. 1434, on oil-engined TS8 chassis, dating from July 1938, had been in service for a year when photographed passing the Aquarium in Brighton. Visible through the archway is an excursion coach on the traditional stance on Madeira Drive.

everyone concerned wants to provide that extra sparkle.

From the earliest days, employees had been provided with facilities for refreshment and recreation, and these improved and enlarged with the opening of each new building. Sporting activities at the depots started humbly with a dart board and went from strength to strength with table tennis, billiards and snooker, football, cricket, road-walking and full-blown athletics meetings being added to the social calendar. A large array of cups, shields and other trophies suitably named after their donors, both Company and 'outside', presented at well-organised presentation nights, turned each programme pamphlet into a roll-call of Southdown officials. Company dances and dinners, presentations of long-service awards, have their origins in this period.

For many years, such events, and much other business concerning the Company, was recorded in a house journal called at first 'Tweenus' and, later, the 'Southdown Chronicle'. This extremely well-designed 'newspaper' had its origins in the example set by Midland Red. R. J. Dallimore, area manager later of Brighton and then Portsmouth depots and a man who had much to do with the running of Southdown's sports and social clubs, bought a Midland Red house-magazine from the conductor of a bus in Birmingham in 1935 and carried the notion of something similar for Southdown back to Portsmouth. The first numbers of 'Tweenus' were compiled, in 1937, at that depot, before production was transferred to Brighton where, over the years, it recorded for posterity the names of hundreds who have contributed to the progress of the Company—among them Cherriman of Hurstpierpoint and Bannister & Evans of Burgess Hill, both of whom transferred to Southdown their stage-carriage services to Hassocks in 1938 and 1940 respectively.

In the final months of peace, when Britain prepared—as A. J. P. Taylor nicely put it—to go "reluctantly to war", Southdown's old ally Wilts and Dorset found itself desperately short of vehicles. The Army was making large extensions to its facilities on Salisbury Plain and the local company couldn't cope. Nearly all the vehicles it normally could have expected to replace, Southdown transferred to Wilts & Dorset—and then found itself short when War Department requisitioning and military production at the vehicle manufacturers became the order of the day. Yet when World War II started Southdown, with its nominal fleet of over 700 cars, was the sixth largest omnibus company in the United Kingdom and its annual revenue had exceeded £1 million.

At the commencement of the war, Douglas Mackenzie devised a scheme to beat the delays caused on the country roads by the black-out. He simply added an extra minute in ten to all the schedules; so that there was a much greater chance of making connections even though a completely overcast sky might greatly reduce the possible speeds. Saloon buses were converted with perimeter seating to take 30 seated and 30 standing passengers. Twelve were allowed to stand downstairs in double-deckers. The Company's war effort included a tax-free loan to the government of £50,000, followed by a War Weapons Week donation of £26,000 subscribed by management and employees. Far more effective than the charming 'Warmington-on-Sea' example, however, was the Company's contribution to home defence.

The Local Defence Volunteer organisation, started in the early summer of 1940, provided the inspiration for Southdown's traffic manager to approach the Home Guard Commander and suggest the setting up within the Company of a separate unit composed of employees and Southdown vehicles. Accordingly, the traffic manager became Lt Col Alexander H. Burman, officer commanding the 12th Sussex (Southdown Motor Transport) Battalion, Home Guard. Over 1,000 employees joined and trained as members of the Royal Army Service Corps—the first battalion to be formed for Motor Transport Service, with its own cooks and clerks included. Until stand-down in December 1944, they put as many as 35 coaches at a time at the disposal of numerous other units during military manoeuvres and field-days at regular intervals throughout the darkest days of the war.

The outbreak of war in 1939 brought an influx of women to act as conductors. The style of hat used was clearly modelled on the contemporary styles of the women's services, notably the Army's ATS, the predecessors of the present WRAC, though some of the girls in this group seem to have been more adept than others in achieving the intended outline. They are standing alongside one of the Leyland Cheetah coaches bought for the Hayling Island services.

The transfer of Southdown to wholly BET ownership occurred in almost the blackest days of the war, so far as bus operation was concerned. Fears for the continuity of oil supplies led the Government to demand that all larger concerns were to convert a proportion of their fleets to producer-gas in place of liquid fuel. A Leyland Titan TD4c dating from 1935, No. 221 (BUF 221), is seen with the gas-producing trailer supplied for this purpose—it ran in this form in 1943-44 along with five similar buses among other vehicles before reversion to petrol was permitted. The combination of gas propulsion with torque converter transmission was rare—this vehicle was later converted to conventional gearbox and later to oil (diesel). The Short Bros body had received the wartime grey roof but was otherwise largely in original form.

Chapter Five: BET's Southdown

For many, 1942 was the turning point of World War II. It was at least the year that a formal alliance of Britain, the United States and the Soviet Union was made against Germany. At home, in the bus world, it was also the year that an alliance was undone. The fourteen-year-old marriage of Tilling and British Electric Traction interests in the United Kingdom—Tilling & British Automobile Traction Ltd—was disparted into its two original components. Southdown was one of the 21 bus companies which found itself in the British Electric Traction Co Ltd group—and in Southdown's case it was a return to the fold. The secretary of BET, Raymond Beddow, joined the Southdown board of directors and Tilling's Thomas Wolsey departed.

Although the Company was obliged to withdraw its remaining express services that September, the absence of holiday visitors was being compensated for by wartime traffic on its stage-carriage services and contract work. Thanks to the high standards achieved by its staff, Southdown had begun its wartime task with rolling stock and buildings in excellent condition. Accordingly, despite being short of both vehicles and personnel—particularly in the maintenance sector—the Company was fulfilling its obligations to the national effort with comparatively few hitches. The strain upon its resources was eased somewhat after 1st January 1943 when a 9.30pm curfew was placed upon bus services by the Regional Transport Officer in order to secure economies of fuel and rubber and relieve the drivers. On Sundays, no buses ran before 1pm.

That month, as though in retaliation for Lt Col Alexander Burman's OBE award, six of the Luftwaffe's new Focke-Wulf 190 fighter aircraft dived in from the sea at an altitude of 50 feet and proceeded to set about a Southdown bus. Midst a fusillade of cannon and machine-gunfire, driver E. J. Hoskin calmly pulled his loaded vehicle up beside a 10ft. bank and frustrated the attack. Curiously and despite a heavy transfer of traffic to the buses as private car owners were forced to give up 'for the duration', such attacks were rare. In 1943, some vehicles serving flat terrain ran on producer gas, brewed up in two-wheeled trailers, as a contribution to diesel fuel economy. It proved difficult to maintain the schedules with such buses, however, and the attempt was abandoned in the summer of 1944. In 1943, the first of 100 Guy Arab utility vehicles arrived to relieve the overworked and intensively repaired Southdown fleet. Many of the early ones had wooden seats which made them seem hard-sprung—but no-one appeared to mind. The hygenic varnished woodwork and the general sparkle of the interiors made a brave effort

Seven of 100 Guy Arab double-deckers supplied under wartime arrangements in 1943-46 had lowbridge bodywork with side-gangway upper-deck, an arrangement used on a proportion of the Leyland Titans supplied in the 1929-39 period. Number 415 (GUF 75) with Weymann-built body was the last bus with this body layout to be supplied—post-war additions to the fleet, including new bodies for existing chassis, were all of the centre-gangway type. With No. 414 of the same type, it had been delivered in December 1944 but both are seen towards the end of their lives eleven years later. Note that it was only possible to match the final digit of fleet and registration numbers—new vehicles were so scarce in wartime that authorities were unwilling to co-operate more than marginally in this way.

to retain the Southdown image. When stage-carriage services in Worthing were drastically reduced, several local drivers took their new Guys over to Portsmouth to help with the preparations for D-Day.

Sadly, the whistling, gear-crashing Guys proved to be the last new vehicles Douglas Mackenzie saw enter the fleet. He died, aged 74, in December 1944, when victory in Europe was well in sight. His beloved Company had come through its severest test with style. Damage to premises had been relatively light. In hard-hit Eastbourne and Portsmouth they seemed to have a charmed life. From its total staff of 2,500, 987 had joined the armed services. Thirty-one had been killed and sixteen taken prisoner. many of the remainder had driven 88,717 passengers around gaps blasted in the Southern Railway network by enemy air activity in addition to the other wartime transport provided by the Company, added to in 1944 by the acquisition of the Bognor-Slindon route of the Silver Queen Bus Service.

Sidney Garcke resigned his chairmanship of Southdown in favour of Raymond Beddow in May 1946, but remained a director until his comparatively early death in October 1948. The imposing, courteous Beddow was to remain chairman of the Company until the Transport Holding Company assumed control in March 1968—a record unlikely to be broken. The long overdue co-ordination agreement with the City of Portsmouth Passenger Transport Department came into effect on 1st July 1946. It embraced operations within the city and the surrounding areas served by Southdown as far afield as Fareham in the west, Petersfield in the north and Emsworth in the east. Mileage and receipts were pooled on a 57 : 43 per cent division, with the Corporation enjoying the larger share. Originally a 20-year agreement, negotiations have since extended the 'Portsmouth Area Joint Transport Services' in modified form to the present day.

Alfred Cannon, a local legend in his lifetime, retired as managing director on 13th June 1947, but retained his seat on the board until his death in June 1952, when a commemorative plaque to his memory was commissioned for display at Southdown headquarters. Area manager at Portsmouth in the difficult years of the war, A. F. R. (Michael) Carling, was appointed general manager following Alfred Cannon's resignation. The man who shepherded his depot through the nightly rigours of the blitz was now to guide Southdown into full peacetime operation and the buoyant years of the early 'fifties. The Company came through a brief period of gloom when severe fuel rationing placed dust sheets over many of its coaches, but put in train plans for new building work at Haywards Heath, Littlehampton, Hassocks and Chichester, together with extensions to the works at Portslade. This was the period when, following nationalisation of the railway companies—including the Southern's one-third share in Southdown—the Thomas Tilling organisation relinquished its United Kingdom bus interests to the state. Arthur Greenwood's 'thirties dream of a common drive towards the social good, now urged on by Herbert Morrison, met stern resistance from BET—and Southdown remained two-thirds un-nationalised.

During the late 'forties several of the old open-top Leyland Titans had the war-time covers (fitted to allow them to remain in all-year round service) removed and ran newly-painted and topless along the seaside routes. In 1950, Portslade Works began lopping tops from war-time Guys and they re-entered service on the scenic run to the Devil's Dyke. Similar vehicles were sent to Eastbourne for the Beachy Head run and to other locations on the coast, until they were replaced by Leyland convertible double-deckers. There was unusual stage-carriage work from Portsmouth in the early 'fifties, as a combination of high unemployment in the city coincided with the development of Esso's refinery at Fawley, a boom in car body-building at Swaything, and of aircraft production at Eastleigh—all in the Southampton area. At one stage Southdown had a bus and its crew based at Hants & Dorset's Eastleigh garage. Back in its own area, the Company purchased the goodwill of Mrs M. Hay-Will's stage-carriage service from Arundel to Burpham.

There was also increased co-operation with another bus-operating neighbour, Maidstone & District Motor Services Ltd. In 1948, an incredibly long run between Brighton and Gravesend was introduced and run jointly with double-deckers. Following the acquisition of the stage-carriage services of Beacon Motor Services (Crowborough) Ltd, controlled by Southdown from 16th September 1949 and fully absorbed in the spring of 1954, and Company control of three services operated by Sargents of East Grinstead, achieved in March 1951, there was further inter-company activity. Maidstone & District gained ground in East Grinstead, but the two companies began to plan a co-ordination scheme which led—on 2nd June 1957—to the establishment of a joint-operation called the 'Heathfield Cycle'. This made all routes in the area, save an infrequent Southdown working—a joint affair, which doubled the number of services operated with Maidstone & District and

With the return of peace, towns such as Eastbourne, prohibited to casual visitors during much of the war period, returned to normality. Here No. 171 (EUF 171), a 1938 Leyland TD5 with Park Royal body, passes Eastbourne station on its way to Brighton on the service 25 route via Lewes. Although free from external scars and remarkably original by most operators' standards in such details as chromium-plated headlamp rims, it was to be rebodied before the end of 1947. It had just passed between two Eastbourne Corporation Titans and the bus pulling into the opposite kerb appears to be one of the Strachan-bodied Corporation Regents. The only post-war vehicle evident is a Southdown Titan PD1 with Park Royal body visible in the distance. The mid-'thirties Austin 'Six' in the foreground was of a type much favoured at the time as a provincial taxi.

The emphasis was very strongly on double-deckers in the post-war period, both in terms of new buses and rebodying of the 'thirties fleet. Here No. 386 (JCD 86), one of the batch of 80 Leyland Titan PD2/1 double-deckers supplied complete with Leyland-built bodywork in 1948, is seen alongside No. 250 (GCD 350), a TD5 dating from October 1939 which had had its original Park Royal body replaced by a new one from the same factory in 1949. They are seen framed by the doorway of Eastbourne bus station in August 1956—the exhortations to queue for the direct service to Brighton and more local routes give an indication of the level of custom anticipated. The dent-free and polished condition of both vehicles at a date when some operators' standards were falling was typical, though perhaps the overall effect was aided by the characteristics of the traditional livery. The PD2 is believed to have been one of those built with non-synchromesh PD1-style gearboxes.

actually reduced some of the fares. At the other end of its area, Southdown acquired the Petersfield-Buriton service when Percy Lambert ceased trading.

Whereas it had been Michael Carling who co-ordinated the detail work for Southdown's running agreement with Portsmouth Corporation, it fell to his successor, Arthur Woodgate—appointed general manager in 1954—to supervise a mileage agreement with the London Transport Executive in 1958, dealing with through services at Crawley New Town and, more importantly, the negotiations which led to the establishment of the 'Brighton Area Transport Services'. Discussions between Brighton Corporation, Brighton Hove & District and Southdown had been taking place for a number of years and agreement in principle was reached in February 1960. An area had been mapped out which included the whole of Brighton, Hove, Portslade, Southwick and Shoreham, including Shoreham Beach. All revenue from stage-carriage services operating within BATS orbit were pooled in the proportion:

Brighton Hove & District 50½%
Southdown Motor Services Ltd 29%
Brighton Corporation Transport 20½%

Part of the revenue from Southdown's cross-boundary routes into Brighton was also accountable and provision was made for services operated wholly within the BATS area to be re-allocated between the three concerns, to give each a fair share of costly services with slower scheduled speeds.

The post-war building programme had been pursued with considerable vigour during the 'fifties: there were major additions to the garages at Eastbourne and Hilsea, new ones at Crawley, Seaford and Moulsecoomb and, in addition to Chichester getting its long-discussed bus station, there were also new ones at Haywards Heath and Lewes. Major contribution to the Company of its architect H. A. F. Spooner, however, was the design and construction-supervision of 'Southdown House' in Freshfield Road, Brighton, the new head office completed in 1964. Southdown also updated its facilities at Hailsham the following year.

Sadly, Arthur Woodgate met an untimely death in 1961, only months after that of his wife. Southdown's unexpected loss was expressed in a beautifully-written tribute by chairman Raymond Beddow, which appeared in the **Southdown Chronicle**, the updated version of the Company's journal. Created, like several other Southdown directors, a CBE in 1956 for his service to the bus industry, Beddow had represented Southdown in the negotiations which led to the area agreements in Portsmouth and Brighton, had taken a personal interest in every aspect of the business, and attended as many Southdown social functions as his numerous commitments would allow. Chairman of the Executive Committee of the BET Federation and for many years a member of the National Council for the Omnibus Industry, he retired in March 1968, still very much a BET man, as the Transport Holding Company assumed control.

Meanwhile, Jim Skyrme was general manager from 1961-66, returned as a director in 1968 and, at the end of the year, succeeded Wilfred Dravers, who had been chairman for nine months. Traffic manager Gerald Duckworth was promoted to general manager in 1966— a creditable performance for, despite being a Lancastrian and in some respects an 'outsider', he had risen to that rank from a relatively humble position within the company. As government pressure upon BET to sell its bus interests to the state mounted, it became clear that Southdown's days as a public company were at last numbered. In January 1968, BET finally agreed to sell its holdings in British bus companies to the Transport Holding Company, set up pending the formation of the National Bus Company. The NBC's re-organisation programme included the transfer to Southdown of the Brighton Hove & District Omnibus Co Ltd in 1969, and Gerald Duckworth became director and general manager of that company also. The initial organisation of the National Bus Company in the Southdown orbit is described in TPC's *Regional History of British Bus Services: South East England*, and the history of BH&D in TPC's *British Bus & Trolleybus Systems, No. 4, Brighton Hove & District* by Southdown's own John Roberts, for many years a BH&D man himself.

The post-war successes of the rugged Leyland PD2 Titans in most parts of the Company area, ensured a continuing patronage for the marque. Here No. 886 (2886 CD), one of the big fleet of Leyland PD3/4 buses with distinctive Northern Counties bodywork that was to be a familiar sight in the Company's territory for over two decades, heads for Brighton in company with one of Brighton Hove & District's open-toppers on the sea front service from Rottingdean. The latter vehicle, an ex-Bristol Tramways and Carriage Co 1939 Bristol K5G with ECW body rebuilt for this service, did not survive until BH&D became part of Southdown.

Southdown greatly expanded its own open-top operations from 1950 and wartime Guy Arab double-deckers were converted to this form at intervals through that decade. Number 420 (GUF 420), a Gardner 6LW-engined example seen here in Pool Valley, was so treated in March 1957, having originally entered service in 1944. Apart from the decapitation, its Northern Counties body retained much of its original form. It was to be among the last to remain in service, being nearly 20 years old when sold.

Partly by force of circumstance, Southdown was not a great builder of bus stations in its earlier years, but in the period after World War II several projects were completed, including the long-discussed Chichester bus station, seen here in the mid-sixties. The single-decker leaving for Littlehampton, No. 101 (101 CUF), was one of the 45 Leyland Leopard PSU3/1RT buses dating from 1963, the first year buses of their 36ft. length were placed in Southdown service. At first, this type seated 51, but were converted in 1965-66 to allow one-man operation, the seating capacity being reduced to 45 to comply with trade union agreements in force at the time. The body design was a product of the BET Federation, though Southdown gave a distinctive look to its version by the waistline treatment at the front. Marshall of Cambridge was a major BET supplier at the time, including all of Southdown's 1963, 1965 and 1967 deliveries.

The validity of the claim to have introduced the first charabanc might be difficult to prove, but Mackenzie's 'slipper' concept, particularly as developed by 1913, had in it the basis of the enclosed motor coach as a species quite distinct from the motor bus, a line of development not taken much further until the mid-twenties. This view of DL 701 clearly shows that almost all windows were of the full-drop type, the only exceptions being at the rear, where dust was a considerable problem on the roads of the period. The Daimler CC chassis is noteworthy for the way in which the top tank was painted but the maker's name on the casting was picked out—in later years, Daimler adopted a deliberate policy of not displaying its name, relying on the distinctive fluted radiator then adopted for identification. This vehicle is believed to be one of those whose chassis were commandeered for military use in 1914.

Chapter Six: Private hire, excursions and 'coach cruising'

When, between March and July 1908, the Sussex Motor Road Car Company provided journeys to Ascot races from the Gloucester Hotel, Brighton, and the Sussex Hotel, Hove, at 10/6 (52½p) return, it set in motion the process by which Southdown was to build its reputation as a purveyor of coach facilities without peer. It was carried further by such private hire work as the military 'motor-dash' to Newhaven Fort undertaken by Worthing Motor Services the following year.

Once installed at 23 Marine Parade, Worthing, Cannon began to build up day, afternoon and evening excursion traffic from the sea front, whilst Mackenzie set about realising his long-felt desire to take members of the public touring to far-off places, the setting up of 'Sussex Tourist Coaches', initially for working excursion traffic from Brighton, together with the delivery of the first 'Silent Knight'-engined Daimler CC chassis gave him the combination for which he had waited. Soon, the Sussex Tourist slogan 'Green Cars Run Everywhere' would have real substance.

There was at first, however, some scepticism. If Mackenzie was sure he now had a reliable vehicle, the potential customers were not. When, in May 1913, he advertised a proposed fortnight's tour from Worthing to the Lake District, he could not get a single booking. It was thought too much of a risk. A group of ladies who listened to his chin-rubbing account of this failure suggested that if he provided a shorter tour lasting "say seven days" they'd provide a nucleus of seven friends—and "the rest is up to you". Several years later, one of those ladies was to become Mrs Eva Mackenzie.

So started the well-documented Devon and Cornwall tour with a full complement of 24 passengers, including Mackenzie, his father and his future wife, in a Daimler CC coach with slipper bodywork of his own design. The journey, which began on 14th June 1913, was full of incident, but each was safely overcome: passengers had to walk up some steep hills; the coach jammed under the entrance arch of the Bude Hotel, Exeter. When the passengers got out, the springs went up and lifted the floor of two ladies' bedrooms above. The rivers Dart and Tamar were crossed by ferry and barge; planks placed over a shingle beach were used to get the coach safely ashore; the brakes failed on Paracombe Hill, but only Mackenzie and the driver noticed; the old-oak scotch was used more than once to stop roll-back; eighteen Lynmouth fishermen were recruited to lift the back end around an impossible corner; when the coach was being driven empty up Combe Hill, a stake protruding from the hedge smashed a side window; and on a narrow road in North Somerset it had to wait half an hour for the daily coach drawn by six horses to come safely through in the opposite direction. "Well, what can they expect," said a small West Country lad, "if they goes about in a bloomin' 'ouse?"

By any standards it was a triumph, and one of the passengers wrote an article for the **Worthing Gazette** called 'Seven Counties in Seven Days' which finished by paraphrasing Macauley: "... The tale's repeated still, How the big Worthing motor drove, Up Countisbury Hill!"

There followed a three-day tour to the

Developments in body design were often of interest in Southdown's early days. This 'charabus' was an attempt to combine the functions of the open charabanc and the closed bus in one vehicle, the 28-seat body being built in 1924 by Dodson on a Daimler CK chassis given the fleet number 35 and the registration number, not 'matching' for once, CD 8569. Somewhat similar opening roof ideas were being tried by other operators at the time, though rarely giving so complete an opening, but this vehicle remained alone in Southdown's fleet, though it survived until 1930. It is believed to have been on solid tyres when new but had received pneumatics by the time this picture was taken in Worthing around 1927/8, though Tilling-Stevens TS3A No. 211 (CD 7711) of 1923 behind, also converted around that period had yet to follow suit.

New Forest in July 1913 with two nights spent at the Stag Hotel, Lyndhurst, and, in September, two coaches labelled 'Aberystwyth' left on a tour to Mid-Wales with a total of 34 passengers and only Duncton Hill in home territory causing any trouble—with the possible exception of a Welsh goose who spread his wings and refused to permit the leading coach to pass. As a result of these successful tours, the 1914 programme started with an 11th May departure for the Lake District, with Mackenzie, Cannon and Percy Lephard among the passengers. This coach sank to its rear wheel hubs on a soft road surface at Ennerdale and in trying to push a projecting stone, Mackenzie brought down at a run an entire dry-stone wall—to the amusement of the owner. When this party returned all available vehicles at Worthing were sent loaded to the Derby. In June two coaches went to Devon and came back in time for Ascot, whilst a Welsh tour returned in July as vehicles were prepared for Goodwood races. There were also 3-day tours to the New Forest and to Kent. Further tours were planned, but the outbreak of war in August 1914 put paid to them. After the Armistice, what vehicles the three-year-old Southdown Company could acquire were needed for stage-carriage work. Additionally, the board of directors did not prove too keen to restart holiday tours. Southdown's first 'coach cruise', as they came to be known throughout the Company's area, was run as an experiment in September 1922 when a Daimler CB charabanc left on a not uneventful trip to North Wales. No tours were run in 1923 or 1924, although a tour to the Lakes was advertised in May and again in July 1923, but did not secure sufficient bookings.

Meanwhile, the acquisition of Arthur Davies' excursion licences at Bognor in 1915 proved to be the first of a long and on-going series of take-overs designed to consolidate and protect Southdown's interest in the field set up by Alfred Cannon. The Southdown coach fleet in the early 'twenties consisted of 36-40hp Tilling-Stevens and Leyland chassis with open canvas-roofed charabanc bodies, relieved from time to time by acquired vehicles of various makes. As early as 1922, experiments included arm-chair type seats with central gangway, thus removing the need for a battery of doors down the nearside. Of the 130 vehicles then owned by Southdown, 50 were charabancs and coaches, and the proportion was increasing rapidly. In the west of the prescribed Southdown area, tours work was considerably strengthened by the purchase of Frank Plater's two companies, South Coast Tourist of Littlehampton and Southsea Tourist Co Ltd in 1924 and 1925 respectively. It was the purchase of the latter, as described in Chapter 4, together with its already established tour organisation which gave Mackenzie his opportunity to re-start his coach-cruises and a few were worked to Devon and the Wye Valley in 1925. The following year, a comprehensive programme of three, seven, ten and fourteen-day tours, booking at Brighton, Worthing and Portsmouth was launched with complete success.

Among the large rival coach concerns bought out by Southdown were J. Poole's

The acquisition of the Southsea Tourist Co Ltd early in 1925, following on that of the associated but smaller South Coast Tourist Co Ltd the previous year, enabled the development of Southdown's touring activities to continue after a period when shortage of vehicles had arrested progress. This 1920 Dennis 3-ton model with 28-seat charabanc body, BK 4530, seen with its original owners, became Southdown No. 360. It was lengthened, fitted with a Dennis E-type engine, pneumatic tyres and a new Harrington 30-seat body at the end of 1926.

Number 480, one of the four rather stubby-looking 'Devon Tourers' of 1928. The Tilling-Stevens Express B9B model was Southdown's standard for charabanc chassis from 1926, most seating 29 or 30 passengers, but this final batch on B9BL chassis had a shorter wheelbase and seats for 20 passengers. The Harrington bodywork was of a transitional type, with a fixed rear dome as well as glass side windows. Full-length running boards were provided but the only entry door on the nearside was at the rear. They remained in service until 1936 and were then sold to Mascot Motors in St. Helier, Jersey, but perhaps the most dramatic development in their lives was being commandeered by the Wehrmacht in 1944 during the occupation of that island—one wonders what the German drivers thought of them.

'Royal Red Coaches', of Hove, whose proprietor was placed in charge of Southdown's Hove office in 1930, and Pott's Coaches of Brighton. In 1932 came the largest purchase of all, the entire coach business of Chapman & Sons (Eastbourne) Ltd. Ironically, one of the horse-brakes which William Chapman had run to the top of Beachy Head had been called 'The Southdown'. He and his sons had turned to motors in 1912 and built up the largest fleet of charabancs in Eastbourne. In World War I, one of their coaches was the first gas-powered vehicle to be seen in Brighton. The family's pioneering spirit flourished after the Armistice and by 1920, tours to Scotland and Wales—and even to France and Switzerland were on the programme. George and William Chapman Jnr continued to expand the business and, by the Easter of 1925, the fleet consisted of 37 Dennis charabancs; tours to Italy and the north and west of Ireland had been added, and over 100,000 passengers were being carried each year.

In 1931, however, coach proprietors everywhere suffered the worst weather for a quarter of a century. It is difficult to understand how one wet summer could have such an effect, but Chapman's, in any case lacking sufficient capital to modernise its fleet, decided to wind up the business. Yet, to a certain extent, a coach company was obliged to hibernate during the winter, whilst those with stage-carriage services had a steady all-year-round return, which supported the tours section during a poor season. Southdown, of course, enjoyed such insurance and, in 1932, a combination of this factor, rising costs, and—above all—the stultifying after-effects of the Road Traffic Act 1930 sealed the fate of Chapmans. Fifty coaches, 44 of which entered the Southdown fleet, the goodwill of the business and certain leasehold premises were sold to the Company, who thereafter enjoyed a patronage built up in Eastbourne over a period of fifty years.

Southdown's coach-cruises now covered the country from Lands End to John O'Groats. Included in their cost was accommodation in first-class hotels and all gratuities. In many instances the more exclusive hotels rejected all coach traffic save that of Southdown—a trend which reached its climax when it became a mark of distinction for an hotel to have a Southdown

Maudslays were predominant in the fleet of Chapman & Sons of Eastbourne by the date it was taken over in March 1932, accounting for 24 out of the 44 vehicles added to Southdown stock. The most numerous type was the four-cylinder normal-control ML4B and No. 333 (JK 1002) was one of five dating from 1930 among those retained until 1936. The 26-seat bodywork of traditional folding-roof type was by Park Royal, which had not only supplied bodies for all but one of the eleven vehicles dating from 1930 but had rebodied fourteen Dennis and two Lancia coaches that year.

Excursion business was steadily building up in the early 'thirties and 'saloon' coaches made it less susceptible to adverse weather. Two of a batch of six Leyland Tiger TS2 models with Harrington 26-seat bodywork supplied in the latter part of 1930 are seen on a half-day excursion, in misty conditions, to Battle about two years later, No. 1021 (UF 6621) being in the centre of the picture. Despite the up-to-date lines of the bodywork, all but one vehicle (sold complete) of the six were rebodied, again by Harrington, in 1936 and remained in the fleet until 1953. The group of vehicles in the background were from the Maidstone & District fleet.

touring coach parked upon its forecourt. By 1934, the Company had become the largest operator of long-distance tours in Great Britain, using finely tuned Leyland Lioness and Cub coaches with Harrington bodywork built to Southdown's own specification. In 1936, they were joined by six beautiful Burlingham-bodied Leyland Tigresses, which became popular with American tourists who booked the tours in advance in the United States or at one of the Company's booking offices opened in London. The drivers were selected from among the best of the long-serving employees and wore chauffeur's uniform with no badges. On their first trips they were taken around the routes by Mackenzie himself, such was his concern that their numerous and diversified duties should be carried out with the utmost professionalism. Relatively high charges were made for the cruises, so that "only passengers of good standing were attracted". Speaking of this with delightful candour in 1934, Cannon opined that "reasonable profits can only accrue when the entire organisation is conceived with the object of catering for luxury coaching as distinct from popular motor tours arranged for passengers belonging to what is commonly known as the 'tripper' class". Fortunately for the Company, a considerable number of the populace reached the acceptable plateau of sophistication as the 'thirties progressed and by the end of the decade the coach cruises were so popular that petrol tankers

Southdown remained faithful to the traditional 'tourer' concept for its extended tours until the mid-thirties. This magnificent Leyland Lioness, No. 314, one of six placed in service in 1933, had a more solid-looking structure up to the top of the window line than earlier versions, but the roof was still of the folding type. Harrington built the body with armchair-like seats for 20 passengers. The chassis was officially of the same LTB1 type as the original Lioness six model of which twelve had been supplied in 1930, but incorporated detail features such as the fully-floating rear-axle and deeper Tiger-style radiator from the contemporary Leyland passenger range. The eighteen vehicles were rebuilt with glass quarter-lights in the late 'thirties and renumbered from 301-up to 1801-up, most surviving until 1951-52.

The facade of Brighton's famous Grand Hotel is seen here in July 1939 with Southdown coaches about to depart on private hire duty. On its board served John Henson Infield, vice-chairman of Southdown Motor Services until his death in 1942. The coaches are 1936 Leyland Tiger TS7 models with Harrington 32-seat bodywork which had just been rebuilt with glass quarter-lights and sliding roofs by Park Royal—they had originally been of the folding-roof type. The leading vehicle, No. 1125 (CCD 725), was to remain in the fleet until 1957, receiving a diesel engine of the pre-war 8.6-litre type in 1950 in place of its original petrol unit.

were outstationed at the more remote spots to refuel the cars. Cruise diaries were introduced and all passengers names entered therein so that there could be immediate communication with them. The coaches cruised at a leisurely average speed of 22 mph and covered no more than 100 miles per day, with no driving undertaken at night, in order that the vehicles were not excessively stressed. The high standards achieved by and with these coaches gradually encompassed all the operations of the Company, including the more prosaic stage-carriage services.

Meanwhile, back on the seafronts from Southsea to Eastbourne, the 'tripper-class' excursions, each carefully advertised by beautifully written advertisements in coloured chalks on blackboards placed beside the appropriate vehicle, continued steadily to build up the destinations and departures. From Southsea,trips to Southampton Docks, the New Forest and Meon Valley were the most popular, whilst farther east, the Devil's Dyke, Winchelsea and Rye attracted considerable custom. In the years prior to World War II, traffic became so heavy that the favoured places upon the stands were allocated on rotation to each proprietor according to the relative size of its fleet. Such was the success of the Company in this quarter, some smaller rivals began labelling their vehicles with fleetnames done in Mackenzie-style script and in a similar colour scheme—until requested to stop by the Traffic Commissioners.

Some signs of a trade recession in hotels nationwide in 1938 had its effect upon bookings that year but, as tension rose in continental Europe thereafter, many travellers used to foreign travel switched to

If imitation is the sincerest form of flattery, Southdown earned its share and although some operators were persuaded to desist, Unique Coaches (Brighton) Ltd paraded some fascinating vehicles in a remarkably Southdown-like livery and condition, except for the choice of a slightly darker green and a diamond-shaped name panel, as late as the early 'fifties. Among them was UF 1813, which had entered service as Southdown No. 413 in 1927. It was a Tilling-Stevens B9B but when rebodied by Harrington in 1934 the chassis appears to have been rebuilt to largely B10B specification. Sold to Ashline of Tonbridge in 1938, it passed to Unique in 1945 and is seen in the traditional Madeira Drive location for a 3/-(15p) excursion in June 1952. It was scrapped two years later— the 'preservation' era was yet to come.

After the wartime cessation of all such activities, coach cruising was resumed in 1947. By that time, the 20-seat Leyland Lionesses had been twice rebuilt, once to the glass quarter-light form in the 1937-1939 period and again as part of post-war rehabilitation when the entrance door was moved from the rear to the centre of the Harrington body. UF 5850 of the original 1930 batch, by then numbered 1802, is seen on a 'Southern Counties' tour—it had received the 'modernised Tiger' style of radiator made by Covrad to update pre-1933 Leylands.

coach cruising in Britain during 1939, and thus provided Southdown with its best year to that point. Even the booking office established at inland East Grinstead the previous year had practically more work than it could handle.

The following summer was a different matter; with the Germans on the French coast opposite, Britain slammed shut its front door, the barbed wire went up on the seafronts, piers were breached to stop unwelcome visitors by steamship and what appeared to be an excursion booking office was much more likely to be a concrete pill-box. Military requisitioning of Southdown coaches had started already in the autumn of 1939, and some soldiers evacuated from Dunkirk and landed at Portsmouth declared that they had "been living in Southdown coaches in France for the last six weeks"—some 24 were converted to ambulances. Others were drafted into the 'Southdown Home Guard', or utilised for carrying munitions workers, troops or ENSA parties far from the leafy lanes of Sussex. At least one Tilling-Stevens became home for a searchlight crew on the clifftop, 'somewhere in Southern England', and others contributed to putting our armies back onto mainland Europe during D-Day preparations in 1944. Thereafter, the Company's coaches provided transport for German prisoners of war working upon road construction and the land.

With the resumption of peace in Europe in May 1945, came the lifting of restrictions upon visitors to the coast. With the barbed wire, if not the larger obstacles removed, waves breaking upon the beach were now met in the opposite direction by hordes of trippers anxious to make up for lost time. With a sadly depleted fleet— thus leading to some mildly unsuitable vehicles being pressed into service—great efforts were made to cope with the weight of traffic. The 1946 service of excursions and tours was thus greatly extended to a proportion well in excess of pre-war figures, and Southdown went back to the Derby and Ascot. Only difficulties in finding adequate hotel accommodation delayed the restart of the coach-cruising programme which actually got going again in time for the 1947 season—20 passengers leaving, fittingly perhaps, for a tour of Devon and Cornwall on 1st May.

Fuel rationing remained, however, and as late as 1948 a government directive led to considerable curtailment of this growing area of business, which in turn presented the Company with manpower problems. It was not until nationalised petrol, stained red to prevent its misuse in private cars, entered the history books, that the upward trend continued. By 1950, however, the year that Southdown started its tours to the continent, some 7,500 passengers per annum were booking on the Company's tours. In 1953, using coaches of a larger seating capacity than previously, Southdown gathered its 'coach-cruising' operation under the banner of 'Beacon Tours', whilst another 'acquired' name, that of 'Triumph Tours' from the Portsmouth area, was adopted in 1960. In 1964, almost 22,000 passengers travelled by Southdown extended tours to many parts of Britain, Ireland and most countries in western Europe. Under contract between 1958 and

Beacon Tours, perpetuating the name of Beacon Motor Services of Crowborough, taken over in 1949, was used as a brand name from 1953. This Leyland Tiger Cub with Beadle centre-entrance bodywork placed in service in 1955 was one of 40 given fleet numbers in the 1000-up series previously allocated to Leyland Tigers by then withdrawn. It is posed at the gateway to Brighton Pavilion——note the Brighton Corporation trolleybus overhead wires visible in the street outside.

1960, Southdown coaches went even farther than that —to Moscow. Large numbers of the passengers carried on these long-distance tours were visitors from America, Australia, Canada, New Zealand and South Africa.

The founding of the National Bus Company led, of course, to the transfer of control of the coach-cruising programme from Southdown to NBC's Central Activities Group; the vehicles involved being repainted in an all-white livery with red and blue lettering. For the best part of ten years, the block letters **SOUTHDOWN,** in red, marked the Company's nominal ownership of these vehicles. Then, from 1980, the spirit of Mackenzie rose up once more, and into service came a coach bedecked in traditional Southdown green, script fleetname and all—and promptly won a class award at that year's British Coach Rally. It has since been joined by others.

In March 1981, in the aftermath of deregulation, the 'Southdown Excursion Club' was introduced to considerably expand the Company's coaching activities. Through a combination of personal service, repeat orders, priority bookings, loyalty discounts and demand-based excursions and mini-tours programmes, the Company set out to increase its excursion and private-hire carryings. The club met with an overwhelming response: the demand for 'Southdown-style' coaching was still there. The summer of 1983 saw a new recruit for such activity enter service: the 'Southdown Diplomat', a twelve-year old coach refitted as 'a luxury palace on wheels', came away from that year's British Coach Rally with four major awards and a great deal of press coverage. Said coaching officer Christine Watts, "It's ideal for business use—whether as a mobile hospitality suite or a conference and dining facility for parties of executives travelling between factories and offices—and it's far more flexible than any executive jet".

Such imaginative use of the Company's resources came at a time when the present upturn in the public use of coach facilities was gaining momentum. For instance, 12,000 passengers travelled in the Company's vehicles to see the various away football matches in Brighton's cup run to the final at Wembley. Cricket was not forgotten either, and a 'Sussex-Link' coach was made available to the Sussex County Cricket team for its Easter tour that year. Not without a sense of humour, the management gained further publicity in the grand manner when a Southdown Leyland Tiger coach was hired for a trip which set a new world record for the Guinness Book of Records. Driven by Ted Elms, the vehicle travelled 320 ft—from Nobles Bar in New Road, Brighton, to the Theatre Royal—with a stop at 'The Volunteer' pub on the way.

Each of these differing examples of the coaching section's activities epitomises its watchwords; reliability, quality and flexibility—words which add up to the good name of Southdown.

Douglas MacKenzie would surely have approved of 'The Southdown Diplomat', complete with lettering in a style not far removed from his ideas of three-quarters of a century earlier. Its collection of four awards at the 1983 British Coach Rally was almost as much a tribute to Portslade Works' skills in restoring a 12-year old vehicle as recognition of its appointments. UUF 329J had started life in April 1971 as one of 25 Leyland Leopard PSU3B/4R models with Plaxton Panorama Elite 32-seat bodywork for long-distance touring. Although originally green, they were repainted in National white coach livery early in life, but No. 1829 reverted to a partially traditional style when converted to the 21-seat executive coach form shown, though the waistband was now gold.

There was another of those echoes of the past which seem to recur in Southdown's rolling stock when five of the present-generation Leyland Tiger coaches placed in service in 1983 were given fleet numbers in a series beginning at 1001, as shown, and thus recalling the original 1001-up series begun with the Company's first Leyland Tigers in 1930. These modern bearers of the name are TRTC11/3 models with Plaxton 50-seat bodywork of the Paramount style introduced the previous year. The livery is a combination of green, white and gold, though the shade of green chosen caused some surprise by being nearer that formerly favoured by Maidstone & District, and thus historically the BAT green, rather than Southdown's traditional apple green.

Early express services were run with charabancs and even up to the beginning of the 'thirties services on the type of route that justified London Coastal Coaches' title were likely to be operated by canvas-roofed vehicles. This scene at the Lupus Street terminal in London shows Tilling-Stevens B9B No. 412 (registered UF 1814 —for once something had gone wrong with the tidiness of the numbering system) dating from April 1927 setting off for what looks like being a wet run to Brighton. Note the Southdown Tilling-Stevens B10 buses and the Leyland Tiger of another operator, possibly West Yorkshire, in the background. Early B9 models had much the same style of radiator as Tilling-Stevens of the petrol-electric era, but this one displayed 'Southdown' in Mackenzie script on the top tank. The vehicle, already looking old fashioned after only three or four years' service, was to be among the 20 which had their original Harrington charabanc bodies replaced by new coach bodies of the same make in 1934-35. Sister vehicle UF 1813 is seen, after rebodying, on page 43; the chassis had also been modernised.

Chapter Seven: Express services

Whilst Douglas Mackenzie was largely responsible for launching Southdown's tours and developing its stage-carriage network, if not its local excursions business, credit for seeing the opening for express coach operations must surely go to Alfred Cannon. Indeed, in some quarters it was said that the cautious Mackenzie was positively sceptical about what, in effect, was a motorised revival of the long-striding stage-coaches of Chaplin and his rivals—particularly since most of the likely routes would offer a direct challenge to the railways.

When in September 1919 the National Union of Railwaymen, under the energetic leadership of James Thomas, took on Lloyd George's government and went on strike against the threat of a wage reduction, several bus companies within range of London saw their opportunity both to make some unexpected extra money and to put notions of such services to the test. Anxious to satisfy the railwaymen before they called upon their alliance with the miners, the government acceeded to their terms. What seemed at the time a total victory for the NUR, however, was in one sense hard-won—for the bus companies found just how easy it was to run limited-stop services to London and, when the strike was over, that there was sufficient demand from the travelling public for them to be set up and expanded upon a permanent basis.

Alfred Cannon had watched with interest the earlier efforts of such operators as Len Turnham of Victoria and William Chapman & Sons of Eastbourne and, in that autumn of 1919, was quick to come to the aid of holidaymakers stranded at the coast, sending several charabanc loads to the capital during the eleven days of the strike. Thereafter, Southdown could ill-afford to ignore the increasingly successful operations of other companies such as Pickfords Ltd, who joined Turnham and Chapman on the Brighton or Eastbourne roads. In 1920, nine companies interested in such activity pooled their express coaching interests and formed London & Coastal Coaches. Among them was Southdown Motor Services Ltd.

Southdown's contribution to the co-ordinated service was actually regularised in 1923 at a time when several other competitors appeared on the Brighton road. This was to lead to the establishment, in April 1925, of London Coastal Coaches Ltd, an organisation specifically designed to apportion mileage, charges and departures, and get the coaches at the London end off their congested roadside stands and into a specially provided coach park. Alfred Cannon was one of the five directors selected to represent the interests of the constituent companies which now included United Automotive Services Ltd and the National Omnibus & Transport Co Ltd. By 1925 Southdown was sending regular departures from Portsmouth, Bognor, Worthing and Brighton to various road-side stands in and around London Victoria.

Finding a site for a coach park was not so easy as expected, and it was not until April 1928 that, as a temporary measure, a two-acre site near Vauxhall Bridge, in Lupus Street, was allocated to the London company by another LCC—the London County Council. In this overcrowded marshalling yard, largely unprotected from the elements, Southdown boasted the tallest notice-board, clearly visible over both charabancs and sprayed mud. By the late 'twenties, competition with operators outside the group had brought the fares from Brighton to London down to 5/-(25p) for a day and 8/6 (42½p) for a period return and, by 1933, there were eleven departures per day on this route alone.

Although a complete 'inverted-fan' of routes to the capital would be established from this framework, from Warsash in the west to Eastbourne, there was to be one route which drew a firm line along the coast itself. In May 1929, Southdown began running between Margate and Bournemouth in conjunction with East Kent and Wilts & Dorset Motor Services Ltd, from whose board Mackenzie had just retired. This was something of a reflexive action, for Elliott Brothers' **Royal Blue** of Bournemouth had pioneered the route, which now saw the sharp competitive running between the 'joint-three' and Royal Blue. However, in March 1932, an agreement was reached between the contestants, whereby Portsmouth would be the dividing place common to them all, the passengers transferring at Southdown's Hyde Park Road

From its introduction to the fleet in 1930, the Leyland Tiger, in successive TS and PS1 variants, was to constitute a virtual monopoly of new express coach purchases for almost 20 years, albeit interrupted by the 1939-45 war. A new fleet number series beginning at 1001 was opened by fifteen TS2 models, of which the first five and one other had bodywork by London Lorries, an unlikely-sounding but briefly very successful contender in the competitive yet fast expanding market for saloon coach bodies, though there was only one other Southdown batch. The low look was fashionable and this picture of No. 1005 when new also shows the two-doorway layout of the 26-seat body, widely favoured at the time, along with the curved-glass front nearside corner window. The London-Portsmouth & Southsea route was an important one from the start, although subject to considerable competition until 1935. Rebodied in 1935, No. 1005 survived until 1953. Though officially limited to 20mph when first introduced, the TS2 with its 6.8-litre six-cylinder petrol engine could achieve 50mph.

depot. Southdown's agreement with Wilts & Dorset was adjusted to give Royal Blue a free reign west of a line from Southampton -Salisbury-Bristol. From 9th May 1932, Royal Blue, Southdown and East Kent launched their joint service between Bournemouth, Southampton, Portsmouth, Brighton and Margate under the title 'South Coast Express'–a name warmingly close to that denied the directors by the Registrar of Companies back in 1915. In fact, all three operators' vehicles were to run through 'on hire' to the 'foreign' sections.

When Chapman & Sons (Eastbourne) Ltd, together with Southern Glideway Coaches Ltd, also of Eastbourne, transferred their licences and vehicles to Southdown in 1932, it represented the Company's first direct acquisition of competitors' express routes. Whereas Chapman's large fleet of worthy Dennis coaches had been engaged on tours and excursions in addition to the London express traffic, Southern Glideway's ten coaches (of six different makes) were concerned solely with its express work. That same year, competition on the Worthing and Brighton services to London eased with the acquisition of the licences held by G. B. Motor Tours Ltd of London and G. F. Wood & Sons Ltd of Steyning–and again in July 1933 when H. J. Sargent's 'East Grinstead Motor Coaches' express interests were merged with those of Southdown. Then, in time for Christmas, another London-based operator was persuaded to give up its express service into Southdown territory: Fairway Coaches Ltd, who'd been sending their Maudslays and Dennis Lancets down to Worthing, handing eight of them over to the Company for disposal.

Meanwhile, on the roads to London from the Portsmouth area, Southdown's Portsea Island-based rivals, North End Motor Coaches (Portsmouth) Ltd, Alexandra Motor Coaches Ltd, T. S. Bruce's 'Imperial Saloon Coaches' and Underwood Express Services Ltd transferred their licences for the London express route to the Company in a flurry of selling which was complete by June 1935. The well-known London-based firm, A. Timpson & Sons Ltd, gave up just its Portsmouth express licences to Southdown during this period, and area manager Chevallier was able to round off a remarkable expansion of express coach working by actually extending operations into Hants & Dorset Motor Services Ltd territory to the west of Portsmouth Harbour. Although Hants & Dorset had established itself at Gosport as early as 1924, it had that same year entered into an agreement with Royal Blue in which it promised not to run any express services, so when G. A. Cross of Gosport wanted to sell his 'Perseverance Motors' express licence for his service to the capital he had to turn to Southdown. Thus the Company gained a route which went from Gosport and Fareham to London up the beautiful Meon Valley. The same applied to Fuger's 'Fareham & Warsash Coaches' licence for similar work, although Warsash passengers for London were later expected to travel to Fareham on the 45 stage-carriage bus and meet Southdown's London coach coming up from Gosport at Hants & Dorset's Fareham bus station.

These acquisitions enhanced the work of the excursion and private-hire department also. For instance, when Portsmouth FC reached the Cup Final at Wembley in April 1934, Southdown and five other local coach companies asked the Traffic Commissioners for permission to run additional coaches to the stadium. Bruce, Cross and Southdown were granted licences at the expense of the others "because they ran a regular service to London". Now that these had been absorbed, proportionally more excursion traffic would come Southdown's way relative to its share of such work. There were now eleven journeys to London each day from Portsmouth alone, some via Putney and Fulham, others via Richmond and Hammersmith, to Victoria Coach Station at 6/-(30p) single, 7/6 (37½p) period return. Certain journeys went beyond Victoria to Kings Cross Coach Station, a short-lived terminus in Euston Road, closed later on the outbreak of World War II.

Stops for refreshment on the London routes had been established at licensed premises on the roads from Portsmouth, Bognor and Eastbourne. So, too, had they at the White Lion on the Worthing route and The Chequers from Brighton. By 1931, however, traffic on the confluent Worthing and Brighton routes had reached such proportion that Southdown erected a half-way-house coach station at County Oak, one mile north of Crawley, at a cost of £10,000. Within four years over a million passengers had used the premises during their break of some ten minutes in the 2 1/2 hour journey. Finding that many of the

Crawley coach station, erected in 1931 to act as a half-way house refreshment stop between London and Brighton, was an immense success—as well as the subject of a famous and lengthy legal battle before alcoholic beverages could be served. The view on the left shows it when newly-built, with contemporary Leyland Tiger and an older Tilling-Stevens. The scene on the right dates from about 20 years later, with Leyland Tigers of the post-war PS1 type and, in third and fifth positions from the left, late 'thirties TS series. The PS1 selection conveys the variety of bodywork purchased in the late 'forties, examples by Harrington, Beadle, ECW, and Park Royal (two) being visible, from left to right.

passengers were asking for alcoholic drinks to go with their snacks, Southdown applied for a liquor licence—a procedure hastened on by the fact that considerably more ladies than men were doing the asking. The first application in 1934, was opposed by the police, the next by a formidable combination of Tamplin's, Portsmouth United and several other breweries with interests on the London road; the Temperance Church, the Congregational Church, the Baptist Church, the Salvation Army and the ultimate in self-denial 'The British Women's Total Abstinence Union'—a not altogether holy alliance. The matter went to court at Horsham in February 1937 when the justices overturned the refusal made twelve months previously to award a 3½ year licence. This, in turn, brought down from the High Court a call for the confirming authority to show cause why a writ of certiorari should not issue. Those who eventually enjoyed "one for the road" in the restaurant and lounge thereafter had little inkling of the complicated and technical legal wrangling which made it possible. Meanwhile this little piece of Southdown history became part of the legal case-book.

Stopping off at the County Oak coach station in the middle 'thirties were fifteen journeys per day each way during the summer on the Brighton run and three on the Worthing with eleven and two respectively in the winter months. Up to 1,080 passengers per day were being carried on these routes. As the Southern Railway Company's electrification of the lines to the Sussex coast was stepped up there had been a slight drop in the traffic which, since the arrangement whereby Southern became a one-third owner of the Southdown shares, had now been stabilised by the inter-availability of rail and coach tickets. In contrast, during this period several sections of the South Coast Express route were unremunerative for the best part of the year.

The acquisition of C. R. Shorter's express licence for the Brighton to Robertsbridge service in 1939 proved to be the last step in the consolidation of such services in the uneasy days before World War II. The war brought difficulties of driving in the black-out and, initially, a rush of mums visiting their evacuated children—until the worst winter for years, followed by war in earnest, began to reduce the programme to a pale shadow of its former-self. Requisition of vehicles, fuel rationing and air-raids, together with the slogan "Is Your Journey Really Necessary?" convinced many people that it was not. Finally, in September 1942, an order issued by the Minister of Transport brought most express services to a halt, and Victoria Coach Station was taken over by the War Office and the National Fire Service. Southdown did its best to compensate with the introduction of limited-stop stage-carriage workings on some sections.

Express coaching began in earnest once again on 22nd March 1946. Large numbers of 'reluctant heroes'—National Servicemen—helped restore the numbers of passengers to healthy levels; a factor which was to lead to 'forces-leave' express services at weekends. Southdown's express service fleet was not allowed to reappear in the somewhat battered state sometimes found elsewhere after the war years. Even the oldest Leyland Tigers were, by exceptional effort, restored to the usual spotless condition. The process was aided by the delivery, from 1947 onwards, of new Leyland coaches with engines which owed much to the technology learned in World War II. Such traffic extended several journeys from Worthing, Brighton and Eastbourne beyond London to Birmingham, particularly during the summer months. There were now services to London from Hayling Island—with a pair of Bedford OB coaches

Entering Victoria Coach Station in the mid-'fifties on a relief working of the Rotherfield-London service acquired from Beacon Motor Services of Crowborough in 1949, is No. 1303 (HUF 303), one of the 25 Leyland Tiger PS1 coaches with Park Royal bodywork added to the fleet in 1947-48. It is in the simplified livery adopted after 1951, all mid-green except for darker green mudguards, but in typical glistening condition, even so. A little touch of modernisation was the addition of the rear hub cover, standard on new Leylands from 1953. Note the semaphore type direction indicator. The traditional-style side destination board was informative even if the front blind display was not quite in the best Southdown tradition.

The Company's full name was originally going to be South Coast Motor Services Ltd, so the use of 'South Coast Express' for the east-west route linking resorts between Margate and Bournemouth was a happy echo, even though the second-choice title has become synonymous with such immense goodwill value over the years. Here Beadle-bodied Leyland Tiger Cub coach No. 1051 of 1956 is seen picking up passengers on a west-bound journey.

purchased in 1948 to augment the pre-war lightweight Leyland Cheetah models—Littlehampton and Rotherfield, together with summer departures from Brighton to London-Oxford-Worcester and to Southampton-Exeter-Torquay-Totnes. There was also some extra work generated by Southdown's link up with Channel Airways' London traffic from Portsmouth Airport. Southdown's interests in such traffic were enhanced in 1954, when the Company's chairman Raymond Beddow became chairman also of London Coastal Coaches Ltd and perhaps even more effectively in 1956 when the latter post was taken up by Southdown's ex-general manager Michael Carling.

An innovation on the Eastbourne-London road on 17th January 1951 had been the introduction of a Leyland Titan double-decked coach. Overweight and upstaged by the large new underfloor-engined saloon coaches, it was relegated to private-hire work after one season. A longer-lasting peculiarity was the retention by Southdown of 'Triumph Coaches' livery. Once part of the Hants & Sussex group, its week-end forces leave express services went as far afield as Liverpool and Leeds. A total of eighteen of the Company's Leyland Tiger Cub and three Leopard coaches were painted in Triumph's cobalt blue and cream livery first applied by an intermediate owner. As well as this Portsmouth-based traffic, Southdown picked up similar work from ex-Hants & Sussex territory at Lee-on-Solent, whilst two further, and last, acquisitions of express licences from rival operators consolidated the series: Unity Coaches of Portsmouth in 1959 and G. F. Graves & Sons (Redhill) Ltd, in 1962.

All the services were given a considerable boost in their competition with the growing threat, the private motor car, in 1961 by the raising of the speed limit for public service vehicles to 40 mph. The 'sixties were noteworthy for the considerable improvements made in the design, construction and performance of the vehicles, together with the high standard of maintenance lavished upon them at regular

Triumph Coaches Ltd had been founded shortly after World War II and built up an extensive group of week-end leave services catering for the large numbers of National Servicemen travelling from Portsmouth to cities in the North of England. After periods in various owners' hands, the share capital was acquired by Southdown in 1957 and the Triumph blue livery and fleetname was applied to a proportion of the coach fleet for several years, mainly to Leyland Tiger Cub models. The last new vehicles were, however, three 1962 Leyland Leopard PSU3/3RT models with Weymann 49-seat bodywork, of which No. 1158 is seen here. They were among the first 36ft. vehicles in the fleet.

Was it an aeroplane, a boat or a motor vehicle? Surprisingly, the Westland SRN-2 hovercraft operated by Southdown for ten days in August 1962 was deemed to be the last-mentioned, and thus an 'express' service between Southsea and the Isle of Wight was provided, even if only briefly. Here passengers are seen disembarking and a noteworthy figure caught passing in front of the photographer's camera, with his own slung over his shoulder, is S. J. B. Skyrme, general manager at the time and later to become Chief Executive of the National Bus Company.

(Foot of page) The Sussex Link concept provides a series of coach services connecting various points between Eastbourne and Havant with London, in a manner reminiscent of the old Southdown express routes, but meeting modern needs. The livery could perhaps be described as 'by National dual-purpose out of traditional Southdown two-tone green', with appropriate lettering. Number 1335 is a Leyland Leopard PSU3E/4 with Plaxton Supreme IV Express body, seating 48 and dating from 1980.

intervals, in particular by Portslade's craftsmen. Noteworthy too for a most unusual experiment carried out by the Company in conjunction with Westland Hovercraft. In what might be called the bravest use of the more abstruse parts of its articles of association ever carried out by a bus company, Southdown actually operated an 'express' service across Spithead from Southsea to the Isle of Wight. Not entirely sure how to go about setting the spray in motion, the Company first applied for licences "to operate hovercraft" from the Air Transport Licensing Board. The Ministry of Aviation, however, had already ruled that a hovercraft was not an aeroplane, but a 'motor vehicle'. This came as something of a surprise for the South Eastern Traffic Commissioners, but from 13th-22nd August 1962, a Westland SRN2 hovercraft bearing the fleetname **Southdown** skimmed back and forth across the waves— ten days which fluttered the dovecotes of the omnibus world. In the event, such services were left to Hovertravel and British Rail 'Seaspeed', but Southdown's reputation for bold innovation had been done no harm.

The late 'sixties proved to be the last years of Southdown's direct control of its old-style express services. Boosted by a further increase of permissible speed to 50 mph, the scurrying green coaches' days were numbered. However, although the National Bus Company was formed in 1968, it was not until 1972 that London Coastal Coaches Ltd changed its name to National Travel (NBC) Ltd. All the constituent companies' express services were placed under the control of NBC's Central Activities Group and standard white National livery was applied to Southdown's contributory coaches.

Throughout the 'seventies, Southdown, of course, provided vehicles for operation upon the routes which it had established, although service provision was determined in liason with National Travel until, in July 1979 and as a sign of things to come, Southdown beat British Rail to a contract to operate the London-Brighton part of a new cross-channel hydrofoil link to Paris. This was awarded to the Company by Jetlink Ferries and its success was all the sweeter when it was learned that the land section to Paris was to be run by SNCF (French Railways). The complete route was London-Gatwick Airport-Brighton operated by the Plaxton-bodied Leyland Leopards painted by Central Workshops, Portslade, in Seajet white to blue livery; Boeing 'Seajet' hydrofoil Brighton-Dieppe and SNCF express Dieppe-Paris.

Then following the enactment, on 6th October, of the Transport Act 1980, National Travel found itself challenged on its express routes by some swashbuckling rivals. Although in the Southdown area it was not as strongly threatened as elsewhere, there was a brief period of independent activity on the London roads from Brighton and Portsmouth but this withered comparatively quickly. However, both economic and political considerations have caused National Express to reappraise its operations in the area. Its own Worthing and Eastbourne routes were now pulled inwards to Brighton whence an hourly non-stop service to London was established. Southdown was now able to provide a series of 'stopping' coach services on what National Travel perceived to be the less strategic routes. Testing the market carefully the Company introduced a three journey 'Solent Link' service on the Southsea to London via Guildford route, but combined most of this service with its S76 'Sussex Link' route from Chichester in 1984. The 'Sussex Link' series of routes, a pleasant reminder of the 'Sussex Tourist, Green Cars Run Everywhere' days of 70 years before, provide green and white liveried coaches running on a growing number of services with widely differing headways. From Eastbourne, Lewes, Brighton, Worthing, Selsey, Littlehampton, Chichester, Hayling and Bracklesham Bay, 'Sussex Link' coaches link the South Coast with London, whilst at the weekends, another runs along the old 'South Coast Express' route from Eastbourne as far west as Havant, where it takes off for Salisbury, Taunton, Exeter and Torquay.

This time-honoured title was officially revived on 28th October 1984, however, when a new limited stop service with an hourly headway commenced running between Brighton and Portsmouth—the centre section of its predecessors.

In the early days of the National Bus Company, its subsidiary operating companies retained their previous liveries. However, supplies of Bristol chassis and ECW bodywork, long-established as the standard practice within the former Tilling group to which their makers had belonged, extended rapidly to ex-BET companies. In part this had already been put in hand before the formal agreement early in 1968 to sell the BET bus interests to the State-owned Transport Holding Company which led to the formation of NBC. There had been widespread recognition of the quality of both Bristol and ECW's products and BET companies, Southdown included, had placed orders soon after they again became available on the open market as the result of a shares deal with Leyland. This Bristol VRT with ECW 70-seat body entered service in 1971 and is seen the following year at Petersfield station on the 42 route from Portsmouth. Note that it bears the traditional-style Southdown fleetname lettering, as slightly modernised in the mid-fifties.

Chapter Eight: NBC's Southdown

Despite diminishing profits, the British Electric Traction Co Ltd retained its British bus holdings until, in January 1968, it agreed under some pressure to sell them to the Transport Holding Company. The National Bus Company, incorporated by statute under the Transport Act, 1968 thus gathered under state control the remainder of the territorial companies, including Southdown Motor Services Ltd.

From 1st April 1972 Southdown was located in NBC's Southern Region. The chief executive of National Bus was S. J. B. Skyrme who had left his general managership of Southdown in 1966 to join the executive staff of BET. Under his leadership, NBC decided to go for a corporate image which would represent to the public a large, reliable and highly experienced organisation. Designer Norman Wilson's package was to include an NBC 'Double N' symbol to be carried on all vehicles and publicity, sales offices, garages and depots, uniforms, badges and flags, bus stops, signs and printed matter, which were themselves to take on a characteristic 'new look'. Stage-carriage buses were to be either red or green with one white band: the treasured Southdown apple green and butter yellow and ornate lettering, including Mackenzie's cursive script, was to go. Southdown buses would be a flat leaf green with white stuck-on lettering.

Almost as though it had seen it coming, the management had pulled out of retirement a 1929 Leyland Titan open-topper, refurbished it in the traditional livery and placed it on a seafront service. The official opening was performed by the Mayors of Brighton and of Hove on 5th July 1971 and the vehicle was crewed by staff in period uniform and the passengers got tickets issued with the aid of a Williamson Bell-Punch machine. The following year, a 1922 Leyland N open-topper retained by Provincial, now also in the NBC fold, was also presented to an admiring public in traditional livery—and one way or another that, and the Mackenzie script, has never quite disappeared from the Southdown scene.

In January 1971, David Deacon had replaced Jim Skyrme as chairman of the Company and of BH&D. February 1972 saw general manager Gerald Duckworth, after 35 years with Southdown, leave to become director of manpower for NBC. He was succeeded briefly by Leonard S. Higgins who, in April 1972, became chief general manager of the so-called 'Channel Coast Companies' (Southdown, Maidstone & District and East Kent) as NBC commenced a revised management structure. Just two months later, Len Higgins went east to become general manager of Maidstone & District and, in preparation for what proved to be a short-lived adjustment to the areas of responsibility for chief general managers, Geoffrey C. Smith replaced him at Brighton in a new grouping which put Southdown and BH&D together with the Isle of Wight operator Southern Vectis.

The standardisation of livery for the now combined fleet of BH&D and Southdown commenced in 1971. Southdown began transferring some of its Leylands to BH&D, new Daimler Fleetlines were shared and some BH&D Bristol VRTs were diverted to the Southdown fleet. Bristol Lodekkas

The now-familiar face of the Leyland National first appeared in the fleet in February 1973, when the vehicle shown was among the initial delivery of a batch of 25 examples, in the then-new NBC livery. They were single-doorway 49-seat buses and were given a new fleet number series beginning at 1.

(Foot of page) Vehicles not yet due for repainting received the NBC corporate image style of fleetname as an interim step. Leyland PD3/4 No. 313 dating from 1966 is seen on the Devil's Dyke Road in 1974.

appeared in Southdown green and cream, marked SOUTHDOWN -BH&D. NBC's green and white livery, which first appeared on new Leyland Nationals at Bognor that year, was adopted for both fleets in 1973. The letters BH&D were removed from the buses in the summer of 1974. Curiously, it was one of the ex-BH&D Lodekkas—a far from typical Southdown vehicle—which was the last double-decker on regular stage-carriage service in the old green and cream of Southdown, some four years later. Two similarly-painted Lodekkas with open tops remain in stock and are brought out in June for the Derby and occasional trips to the Devil's Dyke.

With the absorption of BH&D, Southdown at last turned front-to-back. Belatedly, compared with most sister companies, it now adopted rear-engined double-deckers as standard. First came the Bristol VRT and then the Leyland Atlantean AN68, the latter to a body design much like that evolved for London Country buses.

During the 'seventies the Company adopted the public relations policy of opening its Central Works at Portslade for a day's demonstration of the skills of the various departments and the sale of any memorabilia it could lay its hands upon. In 1974, 4,000 people visited the works; on the workshop's 50th anniversary in 1978 the figure was 7,000. An annual event of no mean proportion had been established.

Following some homework by operations manager Brian Hirst, the old 31 route—the 51-mile Brighton to Portsmouth service—got a new number (700) and a complete face-lift. Launched in January 1975 with its identity writ-large along the length of the vehicles, the 'Coastliner' as it was now called was a speeded-up limited-stop service of rear-engined double-deckers. The first vehicle carried a message of goodwill from the Mayor of Brighton to the Lord Mayor of Portsmouth who sent a reply in the return vehicle. The theme and introductory mayorial salutations were continued with the 'Solenteer', a through limited-stop service from Portsmouth to Southampton via Fareham, in conjunction with Hants & Dorset, which started on 7th March 1976—the first time that such a public stage-carriage exchange had taken place in 50 years. The following spring 'The Regency Route' in harness with Maidstone & District from Brighton to Tunbridge Wells further extended Hirst's distinctly railway-flavoured idea. Several other innovations took place at this time, including the 'Timesaver' limited stop services from the suburbs into Brighton town centre, services from Kirdford to Brighton and to Chichester, a Lewes and district circular and 'Midbus'— a new Seaford town service.

The 'Channel Coast' division was restored in July 1975, when Geoffrey Smith followed Gerald Duckworth to NBC headquarters. Len Higgins resumed the responsibilities he held three years previously and John Birks became general manager. Throughout this period there were numerous changes of management as NBC strove to establish a happy balance at national and local level. In January 1977, John Birks returned to Midland Red as general manager of that firm and was replaced by Michael Sedgley whose appointment was to prove considerably more indelible than had those of his NBC predecessors. On the same day, Irwin Dalton became regional director in place of Frank Pointon who was now promoted to deputy chief executive NBC and, following a September 1977 re-organisation which placed Southdown together with seven other operating companies in the newly-created South Eastern Region of NBC, under Irwin Dalton, previous divisional director Len Higgins was promoted to the post of group executive and consultancy services NBC. Jim Skyrme had retired at the end of December 1976 and singled out Southdown for a memorable farewell 'one-liner' on the standards and courtesy set by its personnel: "It had to be seen to be believed"—enough said!

One of the excuses found for giving its traditional livery another airing was Southdown's participation in a Community Bus Scheme devised by NBC and supported by East Sussex County Council and the Sussex Rural Community Council. The Cuckmere Community Bus project was launched in the winter of 1976-77 with two Ford Transit 15-seaters on a self-help basis. Under the scheme, Southdown provided professional guidance in driver training to PSV standard, advice on licensing, marketing and publicity, vehicle maintenance—and green and cream livery. The villagers, meanwhile, were to do the driving and collect the money over some 340 miles per week. Southdown's own fleet in 1977 was to be enhanced by the latest addition to the open-topped stable—Bristol VR double-deckers, which would eventually replace the Leyland Titan PD3s on such time-honoured services as that from Rottingdean to the Devils Dyke. Meanwhile, entering the world of aesthetics and enthusiasts at one stroke, the SMS publicity and public relations officer got together with the draftsmen at Portslade Works to produce an annual trophy— a polished piston and con-rod—to be presented for

Preserved vehicles kept the traditional livery in use during the early 'strict' years of the NBC corporate image era. Number 135 (CD 7045), a Leyland N with chassis dating from 1922 and Short Bros 51-seat open-top body fitted in 1928, owed its survival to being one of six basically similar Leylands purchased as a stop-gap by the Gosport & Fareham Omnibus Co in 1934 and continuing in service through the war years, being partially restored, initially as a curiosity, and then being superbly rebuilt to 1928 condition after return to Southdown in 1970. Initially the vehicle had been a charabanc on solid tyres. It is seen after taking part in the annual London-Brighton Historic Commercial Vehicle run in 1977.

(Foot of page) Among new ways of tackling the rural bus problem was the Cuckmere Community Bus project, launched in the winter of 1976-77, under which two Ford Transit buses were operated by volunteer drivers drawn from the villages in the Cuckmere valley, a few miles west of Eastbourne. OHC 68M is seen in Alfriston.

'best road vehicle' at the Bluebell Railway Vintage Day.

That summer, Southdown staff at Portsmouth acted as hosts for the crews of 100 additional buses from NBC subsidiaries drafted into the city as part of the local municipality's arrangements for a park-and-ride scheme during the week of the Queen's Silver Jubilee fleet review. Rain considerably dampened the proceedings, but Southdown organisational ability was well demonstrated. The following April at Hove, devotion beyond the call of duty was much in evidence when Southdown staff braved 40ft. flames to drive blackened and charred buses to safety from a blazing Conway Street garage in the early hours of the morning. Nine buses were completely destroyed and five badly burned in the smoke-filled interior. Among those driving the buses out, despite the dangers of explosion and hidden inspection pits, were the men who were to take their buses out on time from 5.50am onwards. Whilst Company officials contemplated the smoking ruins, some passengers out of range of the garage wondered why they were travelling to work on open-toppers meant for scenic work, and relief vehicles were coming in from as far afield as Portsmouth and Eastbourne. All those who helped that day received a personal letter of thanks from the Company.

Another kind of letter introduced that year stemmed from an idea which traffic manager David Bending got from his parish magazine. Local organisations were given the facility to produce their own newsletters or information bulletins inside personalised front covers. The back cover carried bus service news and timetables relevant to the area in which the organisation is based, throughout the Company's 2,500 square mile operating area. Within six months some groups were using as many as 6,000 copies. Further steps to bring Southdown's services to the notice of the community were to involve the local divisional manager for the Brighton area, and members of his staff in taking display material into local factory canteens. The factories also began to relay to their employees the service bulletins by Southdown on Radio Brighton.

In order to test just what it was that the average citizen wanted from his local bus company, Southdown was one of the twelve NBC companies which undertook surveys of its area for the 'Market Analysis Project'. Its participation was launched by Southern Television coverage.

Southdown responded to its newly-gathered information by either adding to, speeding up or reducing its headways in various localities, rather than dividing up its territory into separately identified semi-entities, which many NBC subsidiaries were obliged to do, and which in turn led them into the kind of dipartition whose ramification have yet to be finalised. At Midhurst, which actually saw a reduction in the number of buses provided, an interesting result was the provision of a bus owned by that seasoned contestant Basil Williams, which donned a label marked 'On Hire to Southdown'.

Two of Southdown's own employees could claim to be 'well-seasoned' in 1978-79 when both joined that very special roll of employees with 50 years service. Harold Gill started as a parcel-boy at Hyde Park Road garage, Portsmouth, and Central Divisional engineer Bob Mustchin who had started as a store-boy at Worthing and has seen service also at Bognor and Chichester. Senior citizens outside the Company were not forgotten either, when from 1st June 1979 every man over 65 and woman over 60 were given access to a half-fare pass covering almost the entire Company area. Southdown had already established its annual event for the young: at Christmas

By the mid-seventies, the Brighton Hove & District fleet had been completely assimilated and what had been the final batch of buses to enter service in the BH&D red and cream livery, even though always owned by Southdown, when repainted reappeared in NBC-style green and white. Here No. 2113, the first of the 1971 batch of fifteen Daimler Fleetline CRL6 models with Northern Counties 71-seat two-doorway bodywork, is seen in May 1977 on a local service, in company with a Brighton Corporation Leyland Atlantean in the latter's blue and white livery. Somewhat ironically, the Corporation had also abandoned red and cream after the replacement of the 1938 agreement with BH&D, which had included use of a common livery, so the red hitherto associated with Brighton local buses disappeared completely.

1977 the first of its 'Santa Claus Special' buses went on the road and has since become an established feature. The Company's own loyal band of enthusiasts were not ignored either. Founded in 1954 as the 'Southdown Spotters Club', the Southdown Enthusiasts Club celebrated its silver jubilee, courtesy of the Company, in some style. Its 25th anniversary management committee meeting was held in the boardroom at Southdown House, the Company provided a souvenir cover for the club's monthly newsheet, and a Portslade Works-prepared headboard to go on any coach which they hired in the future. On its own account, the Company reflexes proved up to the moment when the unfortunate vessel *Athena B* ran aground near Brighton's Palace Pier. Out popped the posters: 'SHIPWRECK EMERGENCY! The best way to see the 'Athena B' is from a Southdown bus...' TO BRIGHTON. There then followed a list of the eight services to use. Other peoples misfortunes: the tumbrils were well patronised—well, why not?

A personalised version of NBC's employee newspaper called *Southdown bus news* appeared in December 1980. Disclosing that there were nearly 2,500 people employed by the Company in one capacity or another, they were referred to for the first time by the newly-coined name 'Southdowners'. Also "More conscious consideration of the needs of **customers"** (was necessary) "as opposed to the needs of Southdown to attract more **passengers"** opined traffic manager Philip Ayres; and everyone did their best to call them that for a considerable period afterward. It may not last as long as Mackenzie's **cars** for **buses** but time will tell. Meanwhile, the editor made the irrefragable comment, "Southdown have done a great deal to be proud about".

In June the following year, the paper was able to announce that Derek Fytche had replaced Irwin Dalton as South East regional director and had thus joined the Southdown board of directors. Irwin Dalton, a director of Southdown since 1977, had been its chairman since August 1979. There now followed a period of considerable rationalisation: diminishing financial support from County Councils in real terms, together with a reduced level of new bus grants from central government and less money generally in 'customers' pockets, made it necessary for Southdown to cut its costs accordingly. Apart from the non-replacement of some personnel and the transfer of others at Crowborough to Maidstone & District Motor Services Ltd, the most noticable savings involved the sale of some historic property. The Bognor bus station and depot closed lock, stock and barrel, all vehicles and operations being transferred to Chichester which fairly burst at the seams in an effort to accommodate the newcomers. The old Beacon garage at Crowborough was sold; activities previously based on Emsworth garage were transferred to the new bus station at Havant; Uckfield garage and bus station were closed, whilst Haywards Heath bus and coach station became headquarters for a church. Such measures were to help stabilize the 1982 trading figures and provide a springboard for the operating surplus which was to follow. Doubtless the press, radio and television coverage also helped when, in 1981, consumerist and authoress Elizabeth Gundrey came to Brighton as part of the publicity scheme for her book *England by Bus*. The Company's marketing officer took part in the interviews on BBC Radio Brighton and Southern Television's 'Day by Day' programme, and the occasion was used to launch Southdown's 'Explorer' bus travel scheme—part of a nationwide effort to get people out and about with some of the new, fast services now being introduced by NBC subsidiaries.

Following the enactment of the Transport Act 1980, which deregularised the issue of licences for many stage-carriage activities and placed the onus of proof upon objectors rather than applicants, Southdown began to offer its customers—and those who had yet to acquire that title—a whole new range of limited stop services in the 700 range. Some island-hopped over existing routes, others broke completely new ground. These are listed on the map inside the rear cover, but one which deserves special mention here is that which scurried along from Brighton through time-honoured Storrington and Pulborough all the way to Winchester (after over half a century's absence) and on to the later centre of Mackenzie and Cannon's other 1915 venture—Salisbury and Wiltshire. This was service 710 which operated all the year round: it was joined in the summer by 720 which took in Worthing, Bognor and Chichester before rejoining the 710 route at Petersfield.

The first charabanc, the first buses on pneumatic tyres, the first long-distance tours, the first bus company to involve itself in hovercraft services are some of the claims made, with varying degrees of truth, on behalf of Southdown and its predecessors. There can be no shadow of doubt whatsoever though about its major triumph in 1981. That year, NBC chairman Lord Shepherd announced the introduction of his scheme for giving recognition to new ideas and practices—the Chairman's Award for Innovation. And, for the successful production of bus drivers' fare chart by computer, that first award went to Southdown Motor Services Ltd—the best of 33 entries. By enabling the Company to respond more quickly to market conditions the micro-computer involved had more than repaid its purchase price within a year. Thereafter, instead of the usual 'double-N' logo, Southdown's buses were decorated with a special commemorative badge somewhat reminiscent of the Citroen trademark.

Rather more extensive decoration was now applied to selected buses. The Company's contribution to the English Tourist Board's 'Maritime England', ageing open-topped Leyland Titan 409 DCD was relaunched at Newhaven Fort as a brightly painted grandstand-cum-hospitality suite. A sister vehicle saw service in Bognor Regis as part of the British Tourist Authority's 'Beautiful Britain' campaign, whilst a Bristol VRT in a yellow, red and blue disguise became the '1066 Bus'. This one took visitors from Eastbourne to the historic castle country of Herstmonceux and Camber, and to the site of that quarrel which decided who would be our upper and middle classes—the Battle of Hastings. Early in 1983, this bus went to Dieppe and Rouen and did its best to get more Normans to come across the Channel. In recognition of the Company's contribution to the economic well-being of East Sussex "with particular reference to the advancement of tourism", Southdown was subsequently presented with East Sussex's 'Europe Flag' by Euro-MP Sir Jack Stewart-Clark.

The Portslade paint-shop artists were also at work applying a contemporary white, yellow and orange livery to eight Bristol VRTs from the Conway Street garage for the 'Mile Oak Shuttle'—the first service in Britain run on a 'total-planning' basis. Together with East Sussex County Council and Brighton Borough Transport, Southdown had engaged in extensive research to discover the principal travelling requirements of residents in Mile Oak and Hollingdean. Shuttle bus customers became eligible for discount facilities at certain shops along the routes selling items ranging from fish and chips to antiques. Southdown launched the Mile Oak route, and Leyland Nationals of Brighton Borough Transport followed with the Hollingdean service, with an average ten minute headway. In the 1983-84 football season, another Bristol VRT reflected the local euphoria when the tide of hope came up the beach at Brighton & Hove Albion FC. As part of the fairytale which almost came true, Portslade labelled this one 'Southdown Backs the Seagulls Home and Away', when the local team fought its way through to Wembley and the FA Cup Final. The fact that it was a convertible open-topper shows both what duty it was expected to undertake and who was going to get the job.

In September 1983, Southdown introduced its 32-page 'Good Bus Guide' in answer to those potential customers who claimed that their lack of use of the buses was because they didn't know where to find out about them. The six-part booklet was graphically of a good standard, the right size for the job and had sections on "which bus to catch", and "where they go"; maps showing where they operate in fourteen major towns; how to make use of 'Busranger' (joint with neighbouring Kent and Sussex operators) and 'Explorer' facilities, half fare passes and travel-cards; information on how to obtain timetables; on where to obtain details of all the facilities offered; and a note on Express services, coach hire and excursions. It is one of those rare publications which seems to get bigger as one looks through its pages, so filled is it with just the right kind of information. There was, as the Company claimed, "no excuse for missing the bus" after that.

In December 1984, it was announced that Southdown's organisation was to be split into four bus divisions and one covering the 69 coaches operated. Each was to be autonomous but, unlike those created when dividing some other NBC companies, would not be separate companies, remaining under the umbrella of Southdown Motor Services Ltd. Each of the four bus divisions would have an approximately equal share of the 481-strong bus and express coach fleet, the names being Hampshire, West Sussex, Brighton & Hove, and East & Mid Sussex. Philip Ayers, Southdown's Traffic Manager, was to be responsible for the coach division and each bus division would have a manager responsible to Michael Sedgley, the General Manager.

A few weeks earlier it had been announced that Portslade Works was to become an autonomous Division under the name of Southdown Engineering Services.

Thus to some degree, the wheel of history seems to have completed a full circle, for the divisional names recall the relative independence of the early days. Operations in Hampshire were largely managed from Portsmouth, West Sussex was in even earlier times the province of the Sussex Motor Road Car Co Ltd and 'Brighton and Hove' evokes the BH&D era. Just what the future may bring as a result of the Government's proposals for major changes in the bus industry's structure and operating climate remains to be seen. Yet Southdown's long tradition of offering something extra above the accepted norm is as solid a basis for progress as in those far-off days when Mackenzie, Cannon and French first began to build what was to be one of the most impressive enterprises in the business.

(Left) Among modern marketing campaigns, the '1066 Bus' was noteworthy in being used to interest 20th-century Normans in the scene of their predecessors' exploits in that historic year by visiting France. A map of the area in yellow with blue-painted 'sea' gave a colourful overall effect to the basically red Bristol VRT.

(Below) The County Rider, jointly developed with East Sussex County Council, combines the duties of a country bus with the ability to carry disabled people, thereby taking over the functions of a Social Services vehicle. This was the 1982 prototype, No. 800, with Reeve Burgess body on Leyland Cub chassis; previous Cub models of the 'thirties were used by Southdown on similar routes serving out-of-the-way villages.

Chapter Nine: Rolling Stock

Worthing Motor Services Ltd tended to set the pattern for Southdown's early development, as well as making the largest contribution to the initial fleet. This Milnes-Daimler, CD 361, was among the fifteen ex-WMS vehicles which passed to Southdown in 1915, although the rear-entrance single-deck bus body, of which the number, 6, can just be seen above the bulb-horn alongside the driver's head, had been replaced by a 20-seat front-entrance coach body by then. Body No. 6 had become a van and saw further use on a Caledon and then on one of the ex-Davies Commers. The chassis of CD 361 had started life as one of the Sussex Motor Road Car Co's double-deckers, as shown on page 13. Note how the style of lettering, particularly the 'O' and 'H', resembled the Southdown 'bus' style as used until the early 'fifties.

During the roaring 'twenties the casual visitor to Southdown country could have been forgiven for believing that the Company was exclusively an operator of tinkly Tilling-Stevens buses, so numerous was the type. National servicemen passing along the coast from the mid 'forties could have thought likewise about the whistling Guys. Thirty years on, and the holiday-maker may have thought it a Bristol preserve. Yet the truth of the matter is that although Southdown has flirted with the products of several other manufacturers, its love-affair with Leyland has lasted throughout its 70-year history. Accordingly, it has never been without representatives of the marque, which far outnumber the others.

Nevertheless, a fascinating collection of vehicle-types, whether tucked away in remote dormy-sheds, held in reserve, acquired but not used or placed in service for all to see, has been built up by Southdown over the years. Alan Townsin tells the story in the following pages.

Southdown's vehicle policy took some time before it developed into the very distinctive pattern that, together with the unmistakable 'Southdown sparkle', was to play a big part in creating the Company's special appeal. The initial amalgamation of 1915 inevitably produced a mixed fleet, compounded by wartime difficulties in supply.

By 1919, however, a fairly typical pattern for a company with both British Automobile Traction and Tilling connections had begun to emerge. Ex-military Daimler chassis, in some cases with typical BAT-style bodywork, were much like those entering service in other associated companies. The Tilling-Stevens Petrol-Electric chassis was another typical group choice. Indeed, many of the double-decker examples placed in service until the mid-'twenties were almost identical to those being supplied to the fleet of Thomas Tilling Ltd itself, operating in London and, more significantly, by its Brighton & Hove Omnibus Section and hence often operating over the same roads.

Even so, right from that same 1919 period, there was also an intake of new Leyland chassis, at first N-types with mixed bodywork, some of it second-hand. Leyland had yet to be the universally familiar make of bus it was later to become, though Southdown's neighbour companies both east and west, doubtless under the common influence of Walter Flexman French, had also chosen similarly. For some years, a mixture of Tilling-Stevens (TS3, TS3A and TS6) and Leyland (N, G7 and SG11) chassis were placed in service, (in both cases with double-deck, charabanc and single-deck bus bodywork) together with a few smaller Vulcan models.

In 1926 a fresh pattern began to emerge, with the newly-introduced Tilling-Stevens B9B model with conventional gearbox as the basis of ten charabanc additions to the fleet. The five Leyland G7 double-deckers supplied that year were to be the last Leylands to arrive until 1929. It was rather ironic that just as Leyland was turning to the Lion model which was generally reckoned

continued on page 60

(Above) Requisitioning of chassis for military use at the beginning of the 1914-18 war left Southdown's predecessors with surplus bodywork, some almost new. Among the variety of new chassis purchased very soon after the formation of the new company in April 1915 was this Straker-Squire COT/5 model, numbered 30 and registered CD 3330, on which a 1914 Beadle 29-seat rear-entrance body from the London & South Coast Haulage fleet was mounted. Of advanced design for the period, with full-width windscreen, it and other ex-LSCH bodies set a standard for the post-war Southdown fleet and this particular body was transferred to an ex-Worthing Motor Services Daimler chassis in 1919, surviving until 1925.

(Above right) A similar wartime combination was this Caledon, also dating from 1915, with Dodson 30-seat body very similar to the 'slipper' versions as built for the Worthing company but in fact previously owned by Brighton, Hove & Preston United, having been fitted to an older Milnes-Daimler. Despite its local popularity, this body-design concept was not taken up for the subsequent Southdown fleet. Caledon's radiator of that period had resemblances to later Albion practice, though there was no connection between the two, apart from their common Glasgow origin.

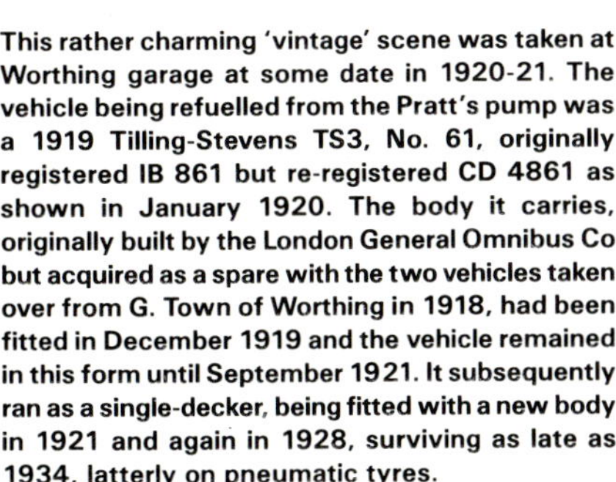

This rather charming 'vintage' scene was taken at Worthing garage at some date in 1920-21. The vehicle being refuelled from the Pratt's pump was a 1919 Tilling-Stevens TS3, No. 61, originally registered IB 861 but re-registered CD 4861 as shown in January 1920. The body it carries, originally built by the London General Omnibus Co but acquired as a spare with the two vehicles taken over from G. Town of Worthing in 1918, had been fitted in December 1919 and the vehicle remained in this form until September 1921. It subsequently ran as a single-decker, being fitted with a new body in 1921 and again in 1928, surviving as late as 1934, latterly on pneumatic tyres.

The military authorities' liking for bus operators' Daimler chassis was reciprocated at the end of the 1914-18 war when large numbers of former Army lorries of this make were purchased for civilian use. As well as a number bought direct in 1919, eleven Daimler Y-types were obtained via other sources in 1920, including No. 24 (CD 5224) in March of that year. The new Dodson 31-seat body was of a style briefly favoured in the immediate post-war period,with a distinctive outline given by the curved roof stiffeners extending forward above the cab—the vehicle retained this body until withdrawal in 1929. It is seen at Worthing on the famous 31 route to Portsmouth.

The 'Southdown sparkle' is evident in this picture, taken at Worthing in the mid-'twenties. The vehicle nearest the camera is of special interest as it was one of three vehicles (two double-deckers and a charabanc from a batch of ten Tilling-Stevens TS3A models delivered in 1922) that were severely damaged in the Bognor garage fire in July 1923, rebuilt and given fresh, more up-to-date, registration numbers. This one, No. 96, originally CD 6896, had become CD 8296 by the date of the photograph. Note the nearside cab door, left swinging open, evidently quite a common practice when drivers left vehicles at this point.

The Vulcan VSD was chosen as a basis for four new vehicles added to the Southdown fleet in 1922, and a further three of the slightly heavier 2T (two-ton) type came in 1923. The VSD model shown came into the fleet with the business of W. G. Dowle, who traded as Summersdale Motor Services, acquired in January 1922. Registered BP 8429, it had been new in 1922, the 21-seat body being by Cutten; the design of its front end is reminiscent of railway Pullman car practice. The vehicle was allocated the number 310 in a series beginning at 301 into which existing Vulcans had been renumbered in 1923, remaining in the fleet until 1928, like most of the type.

Three almost new Ford Model T buses came into the fleet from Shore's Royal Blue Services based in Bognor, purchased in November 1923. The 18-seat bodywork was more substantial-looking than that of most buses based on this world-famous chassis, BP 9685 started another short series of fleet numbers at 331. These vehicles were sold within a year.

Among the more exotic of the mixed bag of vehicles taken over with acquired businesses was this 1922 De Dion 19-seat charabanc from the Cavendish Coach and Car Co of Eastbourne and added to the fleet in January 1925. This famous French maker, more correctly known as De Dion-Bouton, built both cars and commercial vehicles in sizeable numbers until the 'twenties although best known for its early cars. This one, XL 1815, was numbered 47 by Southdown and remained in the fleet for nearly two years.

Forward-control vehicles— those with driving position alongside rather than behind the engine—were not chosen for new additions to the Southdown fleet until 1925, and even then further normal control vehicles were on order. Number 228 (CD 9228), seen here in Madeira Drive, Brighton, in June of that year soon after delivery, is bearing route boards for service 31 to Portsmouth. It was one of eleven Tilling-Stevens TS6 models with Tilling double-deck bodywork plus three single-deckers, but there were also nine (including three charabancs) of the previous TS3A bonneted type, and a further ten double-deckers the following year. Moreover both types had 51-seat bodywork, though the TS6's design permitted a more spacious layout. Most of the batch were traded-in against new Tilling-Stevens single-deckers in 1930.

Similarly, the Leyland SG series of models, the initial 'S' signifying side-type (another term for forward-control), only appeared in Southdown's new vehicle intake in 1925, when seven SG11 models with Tilling 35-seat single-deck bodies entered service; bonneted G7 models with double-deck bodies were to be favoured again in 1926, as well as being a basis for charabanc bodywork in 1925. An SG11, No. 187 (CD 9817) is seen in Uckfield in August 1926. Although fitted with pneumatic tyres around 1927-28, like the rest of the 1925 new vehicle intake, it too was sold in 1930.

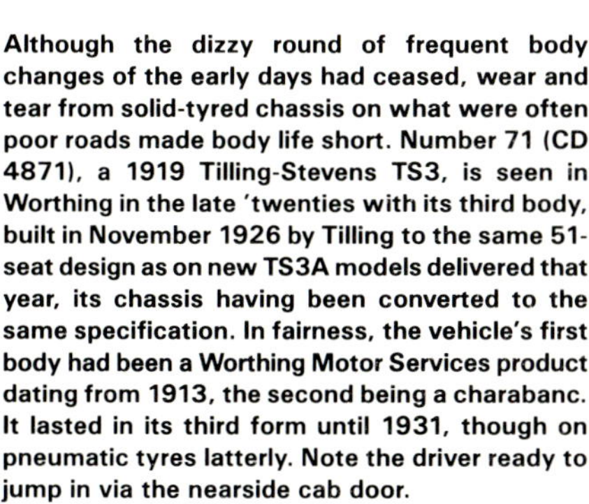
Although the dizzy round of frequent body changes of the early days had ceased, wear and tear from solid-tyred chassis on what were often poor roads made body life short. Number 71 (CD 4871), a 1919 Tilling-Stevens TS3, is seen in Worthing in the late 'twenties with its third body, built in November 1926 by Tilling to the same 51-seat design as on new TS3A models delivered that year, its chassis having been converted to the same specification. In fairness, the vehicle's first body had been a Worthing Motor Services product dating from 1913, the second being a charabanc. It lasted in its third form until 1931, though on pneumatic tyres latterly. Note the driver ready to jump in via the nearside cab door.

its most successful yet, Southdown temporarily deserted the make. Indeed, the Leyland Lion was a model that never appeared in the fleet in any of its successive versions, in contrast to its popularity elsewhere. The 1927 and 1928 new-vehicle intake for the rapidly growing fleet consisted largely of Tilling-Stevens models—more B9B charabancs and a fleet of single-deck buses on the new lower-built forward-control B10A2 chassis. However, there were also some small Dennis 30cwt and G-type buses and coaches, and a solitary Associated Daimler six-wheel open-top double-decker of the 802 type, joined briefly by a second similar vehicle from the Maidstone & District fleet, though both were demonstrators only to remain with the company very briefly.

Leyland returned to favour in 1929, but at first only as a supplier of double-deckers. A batch of 23 of the recently-introduced Titan TD1 model with Brush open-top bodywork supplied early in the year was followed some months later by 22 with standard Leyland-built bodywork of the lowbridge type—the Company's first closed-top double-deckers. However, Tilling-Stevens supplied the 39 single-deck buses and all but one of the 19 charabancs placed in service that year, the exception being an 18-seat Dennis coach—there had also been one Tilling-Stevens open-topped double-decker earlier in the year but it was soon overshadowed by the Titans.

A. F. R. Carling, general manager of Southdown from 1947 to 1954, has described the TD1 as a major development not only in bus design itself but in the way in which it made possible much more widespread use of the double-decker. So far as Southdown's own fleet was concerned, not only were 32 further Leyland-bodied Titans added in 1930 but the refinement and performance of their Leyland six-cylinder engines recognised by the choice of the similarly-powered Tiger chassis for 28 coaches for express services and Lioness bonneted model for eleven touring coaches. The Tilling-Stevens connection with Thomas Tilling Ltd was broken that year and although 32 more B10A2 buses and 46 bonneted B10B2 coaches were added to the fleet, they were the last major intake. Just half a dozen more B10A2 buses arrived in 1931 as well as three of the six-cylinder C60A7 model, the former with 26-seat metal-framed bus bodies by Short Bros and the latter coaches making a last challenge to the Tigers, but Leyland was beginning a long era of dominance.

Southdown's high standards extended to body specifications and the Mackenzie tradition was particularly evident in the development of coaches—the Tilling-Stevens normal-control models of the late 'twenties having more than a little of the look of a contemporary high-grade touring car. Thomas Harrington of Hove was already well established as a major supplier both of coach bodywork and single-deck buses while others for whom Southdown was a regular customer were Short Bros of Rochester, more on the bus side.

continued on page 65

The development of vehicles for pleasure travel began to accelerate again in the mid-'twenties. Pneumatic tyres could now be used on heavier vehicles, though at first sometimes only on the front axles. It is doubtful whether many of the passengers on No. 67 (CD 4867), another of the 1919 batch of Tilling-Stevens TS3, obtained much benefit from this partial conversion. It had retained its original Harrington 32-seat charabanc body but became a TS3A double-decker in October 1926.

This 1919 Leyland N-type, No. 104 (CD 3534—the registration number having been transferred from a Straker-Squire) had received a new Harrington 30-seat charabanc body in 1923. It was fitted in 1924 with what sports car enthusiasts would nowadays call a hard-top and further rebuilt to the rear-entrance layout shown in this display at Eastbourne in October 1926. It had also received pneumatics all round—accommodation for the spare being a new problem. The renewed use of the word 'coach' is noteworthy.

Despite the abandonment of the multi-door layout and the adoption of glass side windows, erecting the hood on this 1926 Tilling-Stevens produced an appearance reminiscent of something from the Western Front in World War I. Number 404 was one of the first batch of a new series of charabanc models on the Express B9B model with conventional gearbox having numbers beginning at 401. Harrington built the bodywork for the batch which was successively altered to modernise the design in 1927 and 1931 before being replaced by new coach bodies in 1935, except for one vehicle sold in 1933—the remainder survived until 1940.

By 1928 a much neater effect had been achieved, as shown (above) by No. 454 (UF 2954), one of a batch of fifteen Tilling-Stevens B9B with Harrington 30-seat bodywork. With a fixed back panel and side windows which dropped into the body sides, these were officially coaches rather than charabancs. The radiator style had now altered to a version with a polished and more rounded top tank. This vehicle was another that found its way to Jersey on withdrawal in 1936 and was commandeered by the German forces in 1944.

The vehicle shown below, No. 442 (UF 2042), was a slightly earlier B9B delivered in 1927 with London Lorries body. It was one of a batch of six, and is shown in a later rebuild from 30-seat charabanc to 29-seat coach, complete with extended 'fixed back' having a luggage carrier on top. Completed by 1934, the vehicle was withdrawn the following year.

Interest in six-wheeled buses was strong in the late 'twenties, and the Associated Daimler Co was clearly hoping for great things from its 802-type double-decker, also known by the London General Omnibus Co type letters LS, standing for 'London Six'. ADC was, in effect, a sales marriage of AEC and Daimler and most 802 models, including the vehicle shown above, had Daimler 35hp six-cylinder 5.7-litre sleeve-valve engines of the type being used in contemporary limousines, and later in the Daimler CF6 bus and coach chassis. Four of the first ten chassis built in 1927 were demonstrators, and Southdown took delivery of 802006 in October of that year, numbering it 28—early numbers tending to be used for 'odd' vehicles by then—and registering it as UF 2638. Short Bros built the 60-seat body, having also built a similar one on the previous chassis, 802005, for Maidstone & District, the only two open-toppers on this model. By November it was in service on the 31 Brighton-Portsmouth route and in February 1928 was joined by the ex-M&D vehicle but both went back to AEC in October of that year, doubtless marking the expiry of a year's trial operation. Only ten more of the type were built, joining two of the earlier ones in the LGOC fleet, and the lack of success probably played its part in the reversion of AEC and Daimler to separate identities. It was to be 30 years before Southdown put larger-capacity buses into service. Number 28 saw further service with the East Surrey Traction Co before conversion to a lorry for the Underground system in London in 1931, becoming LS13 and being scrapped by 1935.

Much more successful were contemporary vehicles at the other end of the size scale. A new 501-up number series was begun in 1927 by a batch of 20 Dennis 30-cwt chassis, all but one having 19-seat bus bodies by Short Bros, then starting a period as a major supplier to Southdown that was to continue until 1935. The use of rear-entrance layout on so small a vehicle was very unusual. The first vehicle, registered UF 1501, is seen here—it remained in service until 1932.

Ex-independent vehicles continued to come into the fleet from time to time, among the more noteworthy being two Unic 14-seat charabancs from C. M. Williams' Pullman Services of Hove in November 1927. The 1921 vehicle shown, XF 4067, became Southdown 43 and both remained in the Southdown fleet for nearly a year. Although of French origin, a somewhat similar but rather lighter chassis was for many years familiar as the basis of many London taxicabs.

(Above) The 1928 batch of five further Dennis 30-cwt buses were of more conventional front-entrance layout but their Short Bros bodywork was metal-framed, rare at the time. The curved glass rear corner windows were also unusual but were to be adopted as standard Southdown practice for single-deck buses until 1939. Note the fabric covering round the chassis dash to allow relative movement between this and the body. Number 526 (UF 3026) remained in service until 1936.

A new generation of full-sized single-deck buses mostly seating 30, 31 or 32 passengers also first appeared in 1928, based on the Tilling-Stevens Express B10A2 chassis, which had a lower frame level than the B9 series. Short Bros, Harrington and, in a few early cases, Tillings shared the body contracts. Ultimately there were 115, delivered over the period up to 1931, though that year's total was only a final six. In a period where bus body design was passing through what was sometimes an unhappy transition in styles, this Southdown standard was notably well-proportioned, and the large and neatly shaped destination box was ahead of most operators' ideas. The first batches were initially numbered 461 upwards but were soon renumbered 601 up and the complete series overflowed after 699 was reached, the last sixteen being 1200-15. Although buses, some were equipped to a slightly more coach-like specification, including No. 639 (UF 4239) with Harrington 30-seat body dating from February 1929 shown (right) on the London-Brighton service. Number 647 (UF 4647), seen below in a Short Bros' official picture of June 1929, shows the radiused window corners introduced on this batch of 20 buses. Although not quite as lively as their name suggested, these four-cylinder vehicles were very reliable—except for the 1928 batch, all remained in service until 1939-40.

The Tilling-Stevens B10A2 was intended as a single-decker but Short Bros built metal-framed open-top double-deck bodies on two, one of which was added to Southdown's fleet. It was given the number 600, 'in front' of the single-deck versions of the same model numbered 601-up, which suggests that it was never regarded as anything more than an interesting experiment. Even so, it remained in stock as a bus until 1946, over ten years after the other vehicle, (which went to the East Surrey Traction Co, registered PK 4244,) had been sold off by London Transport as ESTC's successors. Number 600, registered UF 4300, had been stored during the 1939-45 war but was to serve a few more years as a tree-lopper before being finally withdrawn in 1950. This bodybuilder's view was taken in February 1929. The seating capacity was 48.

The double-deck fleet was given a much firmer step forward a few months later with the arrival of 23 of the Leyland Titan TD1 model. This had been in production for about a year and the measure of its success can be judged by the fact that Southdown's first chassis was number 70465 in a series starting at 70001. Despite the tramcar-like solidity of the 51-seat Brush bodywork, the absence of a top cover allowed the unladen weight to be kept down to 5 tons 11 ½ cwt. The 6.8-litre engine accordingly gave a sprightliness of performance well up to contemporary single-deck standards (where the 5.1-litre capacity of the B10A2's engine was typical), quite apart from the Titan's benefit of six-cylinder smoothness. This bodybuilder's view of No. 804 (UF 4804) acts as a reminder that these vehicles originally had the early type of Titan radiator standard until mid-1929.

Southdown's first covered-top double-deckers appeared in October 1929 when this photograph of the first of the batch of six, No. 824 (UF 5424), was taken at Leyland, where both the Titan TD1 chassis and the bodywork had been built. The latter was of Leyland's standard side-gangway lowbridge type, complete with the enclosed staircase which had only been introduced about two months earlier and was at first an optional feature. Southdown, like most operators of the type, was to specify enclosed stairs on all later examples. Another dozen similar buses followed before the end of 1929 and 24 more early in 1930, all of 48-seat capacity. The era of the Titan had arrived, and with weight only a few cwt more than the open-top version, the nimbleness of the model allowed double-deck operation wherever there was headroom for the unpretentious 13ft. overall height. Southdown was one of few operators to specify its own larger destination box rather than the more modest size standard on this body.

In 1930, there was a brief overlap between the end of the era when a Southdown coach was likely to be a normal-control Tilling-Stevens and the new Leyland Tiger regime. Among the last batches of the 'low radiator' Express B10B2 models were seventeen with 30-seat bodies built by Short Bros rather than the more usual Harrington, and No. 761 is seen here acting as a mobile bandstand for a road safety display in which an Austin Seven is being carried on what appears to be a Saurer lorry. The front roof panel and destination box had been added, modernising the appearance—even so, all but five of the 34 B10B2 coaches dating from 1930 were withdrawn in 1937, the exceptions having been rebodied in 1934.

As the 'thirties began, this involvement blossomed into increasingly distinctive body designs, several also being used by Wilts & Dorset, whose vehicles were often ordered via Southdown as a consequence of the continuing presence of Alfred Cannon on the Wilts & Dorset board through the decade. The 1931 intake of some 55 Titans had Short Bros bodywork of that builder's own characteristic contemporary outline but Southdown gave a 'special' air by specifying opening roofs. Another 32 Tigers joined the coach fleet that year and while the 1932 intake of seventeen Titans (eleven of the last TD1 models and six TD2) and a dozen Tiger TS4 was more modest, the characteristic designs were maintained.

A final six Tilling-Stevens arrived in 1933, in the form of B39A6 models, the successor to the B10 series, with lightweight Short Bros 26-seat bus bodies for the Hayling Island services. Despite attempts with demonstrator double-deckers, by AEC with a Regent in 1930-31 and by Tilling-Stevens with an E60A6 in 1932-35, Leyland had virtually a clean sweep of Southdown business for new vehicles in the remainder of the 'thirties.

This did not mean a lack of variety, however, four models being represented in the 1933 orders for a total of 30 vehicles, for example, the TD2, TS4 and Lioness batches being joined by Southdown's first pair of the small Cub models. In 1934, the new generation TD3 double-deckers and TS6 coaches were joined by the first pair of Tiger six-wheelers in the fleet for the Beachy Head service. Four of the sixteen Titans were of sub-variant TD3c with 'almost automatic' torque converter transmission and eight had oil (diesel) engines, the first in the fleet.

Southdown, like most of the major English bus companies, had taken a cautious view of the oil engine and a first move in this direction in 1934 was about par for such concerns. However, most companies accepted the Leyland 8.6-litre oil engine, smoother-running than most, for bus work for subsequent orders but Southdown reverted to petrol for all its 1935 contracts, including 39 Titans of the TD4 series and a new fleet of 24 Tiger TS7 single-deck buses (the first major intake of single-deckers since the Tilling-Stevens fleet of the 1928-30 period) as well as 24 TS7 coaches and a couple more Tiger six-wheelers.

The last half-dozen petrol TD4 models entered service in 1936, along with another six TS7 buses and a big intake of 47 TS7 coaches, all similarly powered. There were a further six Titans supplied with oil engines that year but these were TD4c with torque converters, as had been some of the previous year's petrol machines. From 1937, with the advent of the TD5 of which 115 were placed in service up to 1939, oil engines and conventional gearboxes became standard for Southdown double-deck orders, and in later years the vehicles dating back to 1934 were gradually altered to this specification.

Body contracts altered somewhat with the disappearance from the bus and coach building industry of Short Bros who were to revert solely to aircraft construction after the completion of the 1935 contracts. Beadle supplied the bodies on the twelve Titans of 1936 but from then on Park Royal, which

continued on page 68

The Leyland Tiger-based Southdown express coach had already moved to a third generation within a year of the first deliveries. Even the first Harrington-bodied examples had not been quite so overtly low-built as the London Lorries version shown on page 47, but the bodywork on numbers 1016-21 as shown on page 42 delivered later in 1930 was slightly taller. In 1931 the two-door layout was abandoned, together with the curved glass front corner window, and seating capacity was increased to 30. The TS2 chassis continued to be favoured, Harrington-bodied No. 1044 (UF 7344) being seen here—by the end of that year, 60 Tiger coaches were in service, the last five on the longer-framed TS1 chassis.

Late in 1930, Short Bros had introduced a stylish double-deck body design with vee-screen outline at the front of the upper deck. Southdown switched to this, deserting Leyland for its later deliveries of Titan TD1 models, starting with four vehicles placed in service at the beginning of 1931. They were of centre-gangway 'highbridge' layout, as had been the last dozen Leyland-bodied examples the previous year. A further 51 Short Bros TD1 models followed later in 1931, followed by a final eleven early in 1932. These latter vehicles were among the last examples of the TD1 to be built, their chassis numbers being in the same series as TD2, TS4 and other models of the new series introduced for 1932. Number 933 (UF 8373), seen here, was the first vehicle of this batch, being sold to Cumberland Motor Services Ltd in 1940, with seven others.

All the 1931 deliveries of Short Bros-bodied Titans had a folding centre section to the roof structure, then quite a fashionable idea. The interior view of No. 878, the first vehicle of the initial four, also shows the original style of single rear window standard on this body design—the split window on the fixed-roof vehicle shown above came in with the opening emergency exit requirements by then in force.

When Tillings severed the connection with Tilling-Stevens in 1930, the associated operating companies' business with the Maidstone-based concern virtually collapsed almost immediately. However, Southdown did take three of the new C60A7 six-cylinder model with Harrington coach bodies in 1931, starting a new 200 series of fleet numbers. The make, now called TSM (the firm's official title having become T.S. Motors Ltd), had adopted a new style of radiator and the overall appearance was arguably more modern than the contemporary Leyland Tiger. The latter was, even so, to hold its place and No. 202 (UF 8032) is seen, despite its smart condition, in its 'for sale' photograph before passing to a South Wales operator—it ended in the United Welsh fleet.

The purchase of the business of Chapman & Sons of Eastbourne in March 1932 brought in some two-year-old Park Royal bodies on Dennis chassis dating from 1925 or earlier. Six Tilling-Stevens B9B chassis built in 1927-28 were purchased from Thames Valley Traction Co Ltd, another company in the TBAT group and the Park Royal bodies transferred to them. Number 490 (MO 9313) was one of the resulting vehicles and the last to remain in the fleet, surviving until 1938. Like most of the six, it seated 22, the chassis being short, in the manner of the 'Devon Tourers'.

(Top right) New vehicle intake for 1932-33 was mainly composed of Leyland Tiger TS4 coaches and Titan TD2 double-deckers. Number 1083 (UF 9783) of the former, with Harrington 32-seat body, is seen in service in 1939—it was to remain in the fleet until 1953.

Although independent businesses continued to be taken over, fewer of their vehicles were taken into stock. Among the last cases where a whole fleet was taken over was that of the Denmead Queen business of F. G. Tanner of Hambledon in March 1935. Among the eleven Thornycroft buses acquired was this A2 model with Wadham 20-seat body dating from 1930, registered TP 9164. It was numbered 43 by Southdown, and then briefly 543 in 1936 when the original number was needed for a new vehicle. It was withdrawn at the end of that year and, with two similar ex-Tanner vehicles, passed to the Newbury & District fleet.

The year 1934 brought three-letter registration numbers and updated chassis designs to the Southdown fleet. Number 973 (AUF 673) was one of a dozen TD3 models with Short Bros bodywork also by then updated in style. The seating capacity remained at the total of 50, with 26 upstairs and 24 down, characteristic of all but the first four Short Bros-bodied Titans of 1931. Eight of these vehicles had oil engines from new, the first in the fleet, and all originally had folding roofs.

Only three new coaches joined the fleet in 1934, but they were to establish the typical Southdown coach outline that would remain virtually unchanged until 1939 deliveries and remain dominant on Southdown express services until well into the post-war period. They were on the Leyland Tiger TS6 chassis with its more compact front-end which, like the corresponding Titan TD3, had been introduced soon after the 1933 Southdown deliveries of TS4 and TD2 models. The Harrington 32-seat bodywork was refined in outline to a form that stood the test of time well. Number 1088, seen here, was to remain in service until 1954, being fitted with an 8.6-litre oil engine in place of the original petrol unit in 1950.

had built six of the 1935 batch of TS7 coaches and the bodies on both normal-control and forward-control Cub buses, increasingly became the builder of Southdown's double-deckers. All of these were of a characteristic Southdown style, though retaining some Short Bros features and having in some cases a hint of Park Royal outlines, and in a similar way the coaches had many obvious Harrington characteristics, even when built by Beadle or Park Royal. Harrington also bodied all the Tiger TS7 and TS8 buses as well as the Cub coaches. Park Royal's lightweight coach bodies on eleven Leyland Cheetah chassis built in 1938-39 for the express service between London and Hayling Island, and for excursions from the holiday camps on the island, had a modified but still distinctive Southdown style.

On the other hand, Burlingham, which secured orders for nine Tiger TS7 coaches in 1936, used virtually one of its own standard styles and even the six Tigress bonneted touring coaches of the same year had obvious detail features of that concern's standard form, though built very much in the Southdown manner for this type of vehicle.

Oil-engined coaches appeared in the form of the 31 TS8 models delivered in 1938 and the 56 TS8 buses of 1938-9 were also oil-engined from new. The 1939 batch of fourteen TS8 chassis sent to Harrington to receive the final pre-war curved-waistline style of Southdown coach body of the pre-war period were also built with oil engines. A major engine-swapping exercise led to almost all the TS8 coaches receiving petrol engines from the 1935-6 TS7 buses, which took the 8.6-litre oil engines thereby released. There appears to be some doubt as to exactly when this occurred, it being reported in some sources as a wartime exercise, but photographic evidence suggests that some of the change-overs may have been made before the 1939 chassis were sent for bodying—several other operators carried out similar exercises.

The only exception to the standardisation on Leyland in the 1934-39 period were related to the take-over in 1938 of the Tramocars concern in Worthing, an order for two Shelvoke & Drewry chassis of a rare rear-engined type being delivered to Southdown, which also ordered two Dennis Falcon models with centre-entrance for the same route.

Wartime stopped the supply of buses to Southdown's normal standards unusually quickly, as a batch of 27 Titan TD7 models with Park Royal bodywork to new versions of the Company's standard highbridge and lowbridge styles was split between three associated companies in 1940. Bearing in mind Brighton Hove & District's difficulties in being allowed to keep buses it was completing in its own workshops at the time, it seems possible that this may not have been Southdown's own choice.

Later in the war years, the need for new vehicles was such that 100 double-deckers were allocated to the Company between 1943 and 1946. All were on Guy Arab chassis, the principal wartime model available, and most of the bodywork came from equally unfamiliar sources. Most had Gardner 5LW five-cylinder engines but nearly a quarter had the 6LW six-cylinder unit and this combination, together with the Northern Counties bodywork allocated on 46 of the total, made a favourable impression.

Southdown was one of the last strongholds of the full-sized bonneted coach and the final six, on Leyland Tigress RLTB3, arrived in June-July 1936. They had bodywork built by Burlingham of Blackpool, a bodybuilder not previously represented in the fleet. Noteworthy were the switch to centre-entrance layout and the form of roof, with only the centre section folding and having curved glass quarter lights. Number 319 (CUF 319) is seen when new—it was to survive with the rest of the batch until 1952, latterly numbered 1819.

The full-width folding hood was not yet dead, however, and No. 1130 (CCD 730), shown below, was one of fourteen out of 38 coaches on Tiger TS7 chassis to basically standard Southdown design placed in service in the period between October 1935 and June 1936 to have the form of roof shown. It was one of 25 of the total built by Harrington, and was one of three not rebuilt to the glass quarter-light form shown on page 43. It went to the War Department in 1940 and did not return to Southdown but returned to passenger service with a Midland independent operator.

A new series of Leyland single-deck buses was started with numbers 1400-up in 1935. These were Leyland Tigers with Harrington bodywork to a new design with some coach characteristics yet quite distinct from the contemporary Southdown coach styles—this photograph graphically conveys that air of something special so typical of the fleet. Number 1402 (BUF 982) was, like the rest of the first 30 vehicles of the type built up to 1936, a petrol-engined TS7 model, though fitted with an 8.6-litre oil engine in wartime. The seating capacity was originally 31, but increased to 32 in 1938—many of the batch, including this one, gave 20 years' service. Note the generous side indicator display.

For lightweight duties, Leyland Cub models were favoured. This SKPZ2 with Park Royal 26-seat body, No. 9 (DUF 9), dated from January 1937.

Double-deckers were equally distinctive. Number 139 (CCD 939) was one of six Titan TD4 models with Beadle 52-seat lowbridge bodywork delivered in May 1936 which were the last petrol-engined double-deckers to enter the fleet. The very shallow roof line was characteristic of Southdown lowbridge buses of this period. Titan fleetnumbers had switched to 100-up in 1935.

TILLING-STEVENS EXPRESS 32-seat Single deck service buses. Upholstered in moquette. Sound mechanical condition and all with C.O.F.

Ready for immediate service . . . . . . **£50** each.

Any trial given . . . . . .
APPLY FOR FULL DETAILS TO . . . .

Southdown Motor Services Ltd.

VICTORIA ROAD, PORTSLADE, BRIGHTON

'Phone - - Portslade 8291

Off with the old, on with the new. The picture of Tilling-Stevens B10A2 No. 606 (UF 3066) at the top of the page was not its official portrait when new but a picture taken in 1936 to advertise direct sale of the first 1928 batches of the type in *'Commercial Motor'*. Even though the value of money has changed, at less than one twentieth of the cost of the cheapest comparable new bus it looked a fair bargain. Similarly, 1930 Leyland Titan TD1 No. 877 (UF 6477) looked pretty sound when photographed shortly before sale to Greenock Motor Services, in this case via a dealer, towards the end of 1938. This was one of the batch with Leyland 'Hybridge' bodies and it had evidently exchanged radiators with one of the 1929 open-toppers being retained in the fleet.

Oil engines became standard for full-sized Southdown buses from 1937 and so all the 115 Titan models placed in service between then and 1939 emitted the characteristic deep roar of the Leyland 8.6-litre engine from new. Number 170 (EUF 170), left, was one of 26 lowbridge Park Royal examples built in the spring of 1938, while No. 205 (EUF 205) was one of five highbridge versions supplied in the summer of that year. The small D-shaped under-staircase window was to remain a Southdown characteristic until 1946. Outswept skirt panels helped to give a stylish appearance though the front-end design and six-bay layout were rather conservative by that date. Both survived until 1961, though rebodied in 1948-49.

The Leyland Cub in various forms was used as a basis for coaches as well as buses, its six-cylinder petrol engine giving a similar degree of refinement as the larger models. Number 30 (CCD 700) was one of six KP3A models of a length normally intended to carry 24-seat bodywork but used in a 20-seat form by Southdown, with generous leg room. Harrington built the bodywork.

When agreement was reached for the purchase of Tramocars Ltd of Worthing in the summer of 1938, two vehicles on order were delivered direct to Southdown. As usual for this operator, they were on S&D (Shelvoke & Drewry) chassis but broke fresh ground in being of a new and very rare rear-engined type. Harrington had been a supplier of bodywork to Tramocars for several years and for these vehicles produced a 26-seat centre-entrance design with front end very like that of this builder's coaches built on side-engined AEC Q chassis in 1934-35. The Tramocar fleet retained its own name, used in addition to the script version of the Southdown fleet-name, and was numbered in a separate T series, the vehicle shown being T16, registered FCD 16. All the S&D vehicles were sold in 1942.

The final fourteen pre-war batch of coaches on Leyland Tiger TS8 chassis dating from June-July 1939 had Harrington 32-seat bodywork of a new subtly modernised style, with gently curving waist and roof outline and curved profile of a subtlety not quite achieved in the post-war period. The chassis were ordered with oil engines but an exchange of engines from these (and some 1938 coaches also oil-engined as delivered) with the petrol engines in the 1935-36 TS7 buses took place. This Harrington photograph inscribed 'August 1939' on the back shows No. 313 (FUF 313) with the emergency top on the Autovac lettered 'Petrol' seeming to imply that some at least of the conversions were almost immediate. Nine of these vehicles were converted to 21-seat touring coaches in 1949-50, evidently being regarded as a better proposition than the harsher-sounding PS1, even though conversion back to 8.6-litre oil was in hand at the time.

The planned 1940 deliveries of Park Royal-bodied Leyland Titan models would have broken new ground in two directions; they were to be on the new TD7 chassis and the front-end styling of the 52-seat bodywork was to be updated, with the by-then widely-favoured convex-curved profile. In the event, they were all diverted to associated companies. GCD 686, which would have been Southdown 286, is seen above as Crosville M124, being one of sixteen to go to that company—Western Welsh received seven and Cumberland four, the last-mentioned being the only lowbridge ones.

The 23 open-top Leyland Titan TD1 buses all received rather ugly temporary canvas tops during the war years. Number 807 is seen soon after being so fitted in December 1942—like most of the batch it reverted to open-top after the war, though not until 1948, being withdrawn in 1950.

The eight surviving covered-top Titan TD1 models and the thirteen remaining TD2 buses were all rebodied in 1943-45, No. 957 (top right) being one of five TD2 models to receive Willowbrook lowbridge 51-seat bodies to wartime utility specification. In typical Southdown fashion, the radiator was still in virtually original 1933 condition, with grille intact, in this post-war view.

Deliveries of Guy Arab double-deckers between 1943 and the early months of 1946 neatly filled the 400-499 series of numbers by then vacant. Although initially there had been no alternative make of double-deckers under wartime restrictions Southdown's favourable experience influenced later orders. Northern Counties, hitherto quite an unfamiliar body-builder in south-east England, was the largest single body supplier and its steel framing, used during the war by special dispensation, also proved durable, much more so than the poor-quality timber in most other utility bodies; No. 418 (left) new in 1944 but rebuilt by Southdown in 1952, was among the last survivors, not being withdrawn until 1963. Number 489, a 1945 Weymann-bodied example (above) was to be among the first withdrawals, its chassis being broken for spares in 1955. Nearly a quarter of Southdown's wartime Arabs had Gardner 6LW six-cylinder engines instead of the more usual 5LW five-cylinder unit, a higher proportion than the national average, justified by the relatively hilly terrain.

The pictures on this page were all taken in the streets surrounding Victoria Coach Station in London over a period of a few week-ends in the summer of 1949. They give some indication of the high standards of presentation of pre-war vehicles on Southdown's express services achieved at a time when many operators' older coaches were clearly showing their age.

(Top left) Leyland Tiger TS4 No. 1084 (UF 9784), seen in company with a two-year-old Leyland PS1, was one of the Company's oldest coaches retaining original bodywork. Dating from 1933, it was still petrol-engined at the time, like all the pre-war coaches shown on this page. The Harrington body had been rebuilt and discreetly modernised the previous year by Portsmouth Aviation; the vehicle was to survive until 1957.

(Top right) This picture of No. 1125 (CCD 725), a 1936 Tiger TS7 with Harrington body, can be directly compared with that of the same vehicle dating from 1939 on page 43—ten more years including the war had made virtually no difference. This was one of the vehicles which had been converted from canvas to 'glass' roof, but No. 1135 (CCD 735) of the same batch visible behind had always been in the conventional express-service form shown—its body was built by Beadle to Southdown specifications. Both were to be converted with Leyland 8.6-litre oil engines in 1950-51, like most of Southdown's surviving pre-war Tiger coaches.

(Second row, left) FUF 505 was one of the second, 1939, batch of lightweight Leyland Cheetah coaches with Park Royal bodywork for the Hayling Island service, by then numbered 605 and seen in company with No. 70 (JCD 370), one of the two Bedford OB Duple Vista coaches bought for the same duty in 1948.

(Second row, right) Number 1155 (CUF 155) was one of the nine Tiger TS7 coaches with Burlingham bodywork dating from 1936. Although different in style to the rest of the pre-war fleet they remained a familiar sight until the mid-fifties.

(Bottom) Heavy traffic on summer week-ends sometimes caused vehicles meant for tour or excursion work to be used on the London services. Number 48 (DUF 48) was a 1937 Leyland Cub KPZ2 with Harrington 20-seat body which remained in the fleet until 1956.

Post-War reconstruction

Southdown had much the same problems in overcoming the after-effects of the war years as most other operators. Yet in a surprisingly short time the Southdown sparkle had returned. What was truly impressive to behold was the way in which the exceptionally high standards were restored not merely to a few vehicles but to the whole fleet. By the time I first began to see Southdown coaches regularly, from 1947-8, I simply **never** saw anything but a smart vehicle— it was fascinating to walk around Victoria Coach Station and nearby streets and see coach after coach in the condition nowadays associated with preserved vehicle restoration standards.

Visits to Brighton and elsewhere showed that the bus fleet was almost as good, though the structural condition of the timber-framed double-deck bodywork, almost all over the life-span of about eight years intended when it was built, had created problems. Rebodying had begun under wartime rules and after the war the programme was extended, with East Lancashire and Northern Counties from the wartime suppliers of bodywork joining Beadle and Park Royal and being further augmented by Saunders, a firm which had inherited something of the character as well as the management of Short Bros.

New vehicles came from Leyland, restored to its position as main chassis supplier, and also supplying double-deck bodywork for the first time since 1930, though the first 25 post-war PD1 chassis for the fleet received an ingeniously 'civilised' version of Park Royal's wartime body. The Tiger PS1 became the basis of the early post-war coach fleet, 125 vehicles dating from 1947 to 1949, with a variety of bodywork. Only Beadle was now prepared to build to Southdown design (incidentally maintaining an old tradition by using the same design for Wilts & Dorset, no longer in any way related to Southdown, and in this case on Bristol L6B chassis). Duple produced a version with Southdown rear-entrance layout but based on its own structure, Park Royal built to East Kent design for both companies while Harrington, surprisingly in view of past associations, simply supplied its contemporary standard front-entrance coach, but without the customary 'fin', as did new supplier Windover. Perhaps the most unexpected vehicles were the first 25, which had Eastern Coach Works bodies of that firm's immediate post-war standard—an ironic time for the first Southdown bodies from this source, as the split of TBAT in 1942 had put ECW and Southdown, hitherto both in the group, into 'opposite' camps.

continued on page 78

In the early post-war period, even the 'tourers' were apt to be pressed into express services. Here UF 8830, one of the Lioness 20-seat Harrington-bodied coaches dating from 1933 and by then renumbered 1816, is seen on arrival at Victoria. It was substantially as rebuilt in 1938-39, with glass quarter lights—a further rebuild in 1946-47 moved the entrance door amidships. The lack of front wheel nut ring marked a very brief lowering of normal standards, soon restored.

Most of the 1930 Leyland Tiger TS2 models rebodied in 1935-36 survived until 1953, including No. 1024 (UF 6924) seen (above) in Portsmouth in 1949. The 1936 Harrington body was one of those with a folding canvas roof.

It wasn't just the coaches that were maintained to a high standard, even when getting on in years. Here Harrington-bodied Leyland Tiger TS8 oil-engined saloon No. 1461 is seen at Pool Valley in Brighton around 1950. Dating from November 1938, this had been one of the vehicles converted to 'perimeter' seating in wartime to allow more standing passengers to be carried, in this case in November 1942, reverting to the normal 32-seat layout in December 1946. It remained in service until 1956. The dark green roof was adopted for single-deckers towards the end of the war and retained for new deliveries in the 'fifties.

The only additions to the fleet in 1946, apart from the last five utility Guys, were 25 double-deckers on the recently-introduced Leyland Titan PD1 chassis. The Park Royal bodywork was finished to peacetime standards and given some deft touches of Southdown flavour, such as the understaircase window and the louvre over the windscreen but was based on the standard Park Royal utility-style body framework, producing its most glamorous direct derivative. They were given fleet numbers 266 upwards originally intended for the TD7 buses diverted elsewhere in 1940, for which they were, in effect, replacements, though two fewer in number. GUF 283 is seen before delivery in original livery with dark green roof, a wartime feature retained until 1947. Like most of the batch, it remained in service until 1963.

One of the thirteen TD4 models which retained its original body was No. 104 [BUF 204] of 1935, as shown above, though its rather angular Short Bros 52-seat body was rebuilt by Saunders in 1946. Significantly, the latter concern's bus bodybuilding activities were established by an ex-Short Bros man. Saunders went on to build nine new 54-seat bodies, No. 183 [EUF 183], a 1938 TD5, being the first to arrive, in June 1947 as shown [right]. These incorporated the characteristic Southdown under-staircase window, a similar vehicle being partially visible on the right of the picture above [the one on the left being one of the 1946 Park Royal-bodied PD1 models].

A major rebodying campaign was put in hand for the Leyland Titan TD3, TD4 and TD5 double-deckers dating from 1934-39. In the early stages this followed on from the wartime rebodying of TD1 and TD2 models, involving small numbers of vehicles, but the work was carried out on a larger scale from 1947, continuing until 1950. Ultimately, 153 out of the Company's 182 chassis of these three types had been rebodied and they then continued in service until 1958-62, some chassis thus completing 25 years' service. Such was the pressure on the bodybuilding industry that the work was split between five concerns, each using their own designs.

[Top] Beadle rebuilt most of its own 1938 batch of lowbridge bodies on TD5 chassis in 1944-46, including No. 198 [EUF 198] seen in Bognor in 1949.

[Above] The first dozen Beadle bodies of the post-war series, dating from 1947-48, were this builder's sloping-front six-bay style. They included No. 120 [BUF 220], a 1934 TD4 shown at South Parade, Southsea, after having been rebodied in August 1947. Seating capacity was to Southdown's post-war standard of 54 [28 up, 26 down].

[Right] East Lancashire was the biggest participant in the programme, contributing 60 bodies. The earlier examples were of 52-seat capacity and built to this builder's design as introduced in wartime, though, as with Beadle, not conforming to the utility specification. Number 114 [BUF 214], a 1935 TD4, was rebodied in August 1947 and is seen at Haywards Heath with No. 1538, a Leyland Royal Tiger also with East Lancashire body dating from 1953.

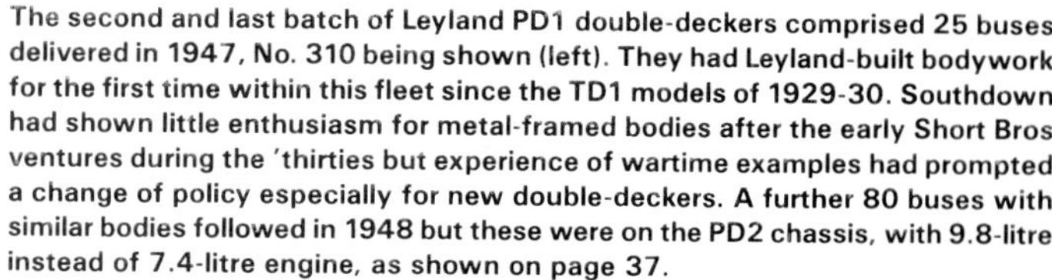

The second and last batch of Leyland PD1 double-deckers comprised 25 buses delivered in 1947, No. 310 being shown (left). They had Leyland-built bodywork for the first time within this fleet since the TD1 models of 1929-30. Southdown had shown little enthusiasm for metal-framed bodies after the early Short Bros ventures during the 'thirties but experience of wartime examples had prompted a change of policy especially for new double-deckers. A further 80 buses with similar bodies followed in 1948 but these were on the PD2 chassis, with 9.8-litre instead of 7.4-litre engine, as shown on page 37.

(Below, left) The 20 Beadle bodies supplied for existing TD-series chassis in 1949-50 were of a revised design based on a five-bay structure and with a more conventional type of cab front. Number 141 (CCD 941), a 1936 TD4, had originally had a lowbridge body by the same builder but was rebodied as shown in May 1949.

(Below) Easily the most curvaceous style of body among those chosen for the rebodying programme was that of the seventeen supplied by Northern Counties in the early months of 1950. Number 257 (GCD 357), had been one of the last TD5 models delivered, originally entering service with Park Royal bodywork in November 1939. Seen here, after rebodying, in Worthing, it was again one of the last of the type to be withdrawn, in 1962.

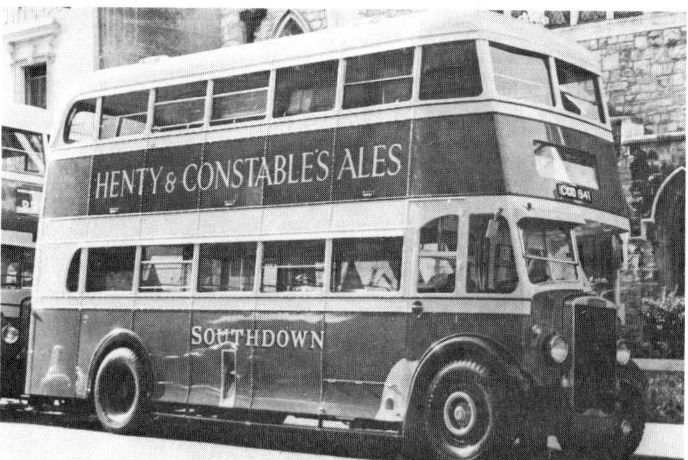

East Lancashire was another bodybuilder to introduce a second style during the Southdown rebodying programme, in this case less rounded in profile than the earlier version. Number 151 (DUF 151) had been the first TD5 in the fleet when new in July 1937 but is seen here as rebodied by East Lancashire in 1949. Alongside it in the Pool Valley terminus in Brighton is No. 205 (EUF 205), the last of the 1938 batch of TD5 models, also rebodied in 1949 but by Park Royal. It was one of 35 vehicles to receive this style of bodywork, basically this builder's metal-framed standard of the period.

Bodywork for the early post-war coach fleet additions was similarly mixed, though the Leyland Tiger PS1 7.4-litre-engined chassis was standardised for all 125 full-sized vehicles—they entered service between February 1947 and November 1949, all but the ECW version being 32-seaters. They remained in service until 1959-62.

(Below) The most faithful rendering of Southdown styling ideas came from Beadle, with a rear-entrance coach not quite matching the graceful lines of the 1939 vehicles but combining distinctive Southdown features in a body structure with straight roof line but gently curving waistline. A total of 23 were supplied between December 1947 and August 1949.

(Top left) The first to arrive had bodywork by Eastern Coach Works, basically to the Tilling group's post-war single-deck design in its 31-seat 'express' form; little but the rear-entrance layout conformed to Southdown's practice. The top-sliding windows evident in this early view of No. 1249 (HCD 449) at Victoria were later replaced by half-drops. Delivery of the 25 was completed by May 1947.

(Top right) Harrington supplied six examples of its standardised front-entrance sloping-pillar curved-waist coach design. The first was of the half-canopy type and the only concession to Southdown standards was a switch to a full-width canopy allowing the normal style of destination display on the remainder. This one was No. 1265 (HUF 5), delivered with the first in April 1947—completion of this small order took until February 1948.

(Second row, left) Park Royal had resumed building rear-entrance coaches for the East Kent Road Car Co Ltd in 1946, using a mildly updated version of a pre-war EK style, and this was accepted as a reasonable approximation of Southdown's standard for twelve coaches supplied in May 1947, including No. 1260 (HCD 860) seen here and another dozen in March 1948.

(Second row, right) Windover, old-established builders of luxury car bodywork, joined the post-war coach trade, using a standardised design represented here by No. 1272 (HUF 272), one of six supplied to Southdown in October-November 1947.

(Right) Duple's interpretation of Southdown's ideas was similar in concept to the Beadle version but incorporated characteristic Duple outlines—the end result nonetheless being attractive in overall effect. Delivery was spread between June 1947 and November 1949, the total building up to 40, making up the largest single group of PS1 coaches.

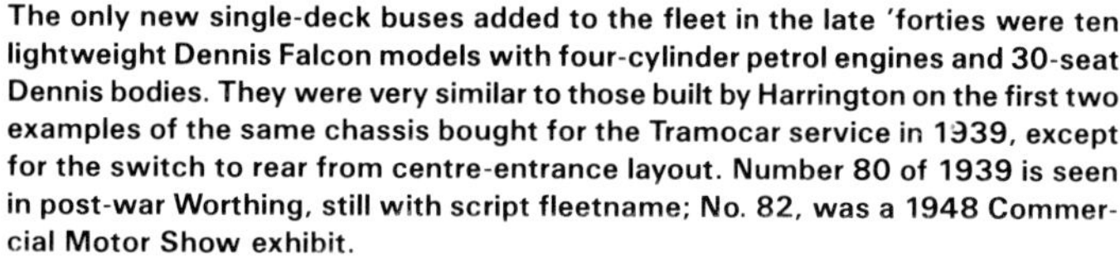

The only new single-deck buses added to the fleet in the late 'forties were ten lightweight Dennis Falcon models with four-cylinder petrol engines and 30-seat Dennis bodies. They were very similar to those built by Harrington on the first two examples of the same chassis bought for the Tramocar service in 1939, except for the switch to rear from centre-entrance layout. Number 80 of 1939 is seen in post-war Worthing, still with script fleetname; No. 82, was a 1948 Commercial Motor Show exhibit.

(Top of page) Another 1948 Show exhibit was No. 502 (JCD 502), one of twelve Guy Arab III models with Gardner 6LW engines and Northern Counties bodywork, but differing from the rest (see page 81) in having a more elaborate body design with a patented ventilation system. It paved the way for (above) the 1950 Northern Counties exhibit, a double-deck coach, No. 700 (KUF 700), on the newly introduced PD2/12 version of the Leyland Titan chassis. This took advantage of the 27ft. by 8ft. dimensions first permitted for double-deck two-axle vehicles in general operation that year. As built, it seated only 44 passengers, being intended for the Eastbourne-London service, but with an unladen weight of 8ton 13cwt 1qr, over a ton more than the PD2/1 buses, its performance on gradients was rather sluggish. Reseated twice, it was tried on bus work and private hire before early withdrawal in 1966. It had been the only new vehicle purchased in 1950, no vintage year for Southdown fleet additions.

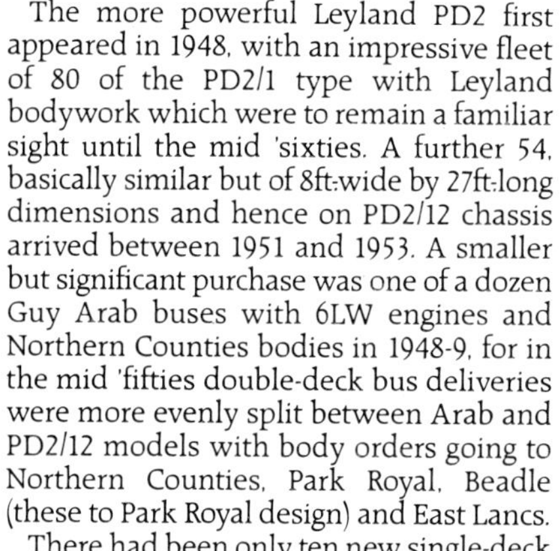

The more powerful Leyland PD2 first appeared in 1948, with an impressive fleet of 80 of the PD2/1 type with Leyland bodywork which were to remain a familiar sight until the mid 'sixties. A further 54, basically similar but of 8ft.-wide by 27ft.-long dimensions and hence on PD2/12 chassis arrived between 1951 and 1953. A smaller but significant purchase was one of a dozen Guy Arab buses with 6LW engines and Northern Counties bodies in 1948-9, for in the mid 'fifties double-deck bus deliveries were more evenly split between Arab and PD2/12 models with body orders going to Northern Counties, Park Royal, Beadle (these to Park Royal design) and East Lancs.

There had been only ten new single-deck buses between 1939 and 1952, these being more lightweight Dennis Falcons for Hayling Island, where there were also a couple of Bedford OB coaches. The arrival of underfloor-engined models brought a new generation of both buses and coaches, the latter incidentally killing a double-deck coach project which never got past the prototype stage. The original Leyland Royal Tiger chassis was favoured between 1951, when the first ten were delivered as touring coaches, and 1953. The first buses maintained the old Southdown rear-entrance tradition but, from 1953, centre-entrances were favoured. It was to be 1959 before the first conversions to front-entrance were made.

continued on page 82

The 30ft.-long underfloor-engined single-decker offered a more promising line of coach development, though Southdown's first ten were modest indeed in passenger capacity, seating a mere 26 in 'two and one' formation as they were intended as touring coaches, taking over from the Lioness and Tigress fleet. They were on the Leyland Royal Tiger chassis, with basically the same O.600 9.8-litre engine as the PD2 and had many other mechanical features in common. Duple's success in supplying PS1 coach bodies was to be followed by further orders for coaches on Royal Tiger chassis, these at first being PSU1/15 models with the vacuum brake system much the same as on contemporary double-deckers. A new numbering series was begun at 800, the vehicle shown being No. 808. Duple named this body design the Ambassador, 20 more with 41-seat layout beginning a new 1600-up series in 1952.

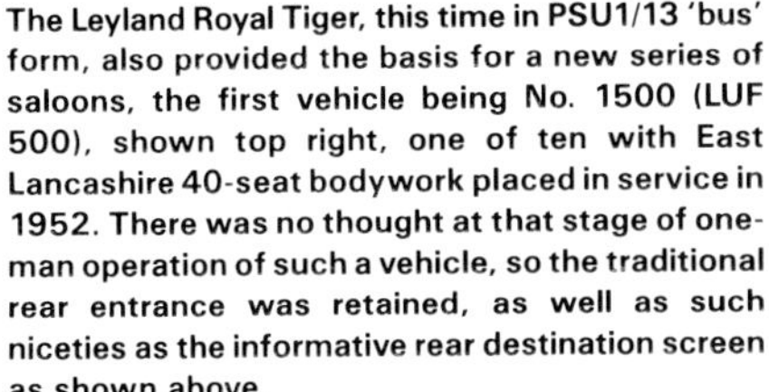

The Leyland Royal Tiger, this time in PSU1/13 'bus' form, also provided the basis for a new series of saloons, the first vehicle being No. 1500 (LUF 500), shown top right, one of ten with East Lancashire 40-seat bodywork placed in service in 1952. There was no thought at that stage of one-man operation of such a vehicle, so the traditional rear entrance was retained, as well as such niceties as the informative rear destination screen as shown above.

The 1952 intake of Royal Tigers was more predominantly of PSU1/15 coaches, however, with the express-service fleet receiving 45, including 25 with Leyland's own 41-seat coach body for this model, including No. 1637 (LUF 637) shown above with boards for the London-Bognor route. They were solidly-built vehicles but an unladen weight of 8tons 6cwt, though not unduly high among early underfloor-engined models, was beginning to cause concern in an increasingly cost-conscious climate. They remained in service until the mid-sixties, latterly mostly in dual-purpose form.

(Above and right) Harrington came back into the picture after a gap of nearly five years with 20 touring coaches also on the PSU1/15 chassis and similar in seating layout to the Duple vehicles of the previous delivery, again offering what could best be described as armchair comfort to 26 passengers apiece. The Harrington version offered rather better vision, aided by use of glass panels above the windscreen which were not, as might have been thought, intended for destination display. Number 818 (LUF 818) is seen in both views.

(Above left and above) Although the arrival of Royal Tigers began to signal the end of the pre-war Tiger bus and coach fleets, major parts of the chassis of 50 of them were to live on for another decade or so. Like several BET companies, Southdown took delivery of Beadle integral coaches using pre-war Leyland units, the first batch of 20 delivered in 1952 being 30ft. long and seating 35 passengers—No. 861 (LCD 861) being typical. The second 30, dating from 1953-54, were more unusual, being 26ft. long 26-seaters, including No. 871 (MCD 871). The running gear, including 8.6-litre oil engines, was from Tiger TS8 models, the chassis being cut to form front and rear sub-frames. The weight of the 30ft. ones were much the same, at about 6¾ tons, as the original pre-war coaches.

(Above) Coronation year, 1953, was marked by Duple by giving this new coach design the name Coronation Ambassador and No. 1647 (MCD 47) was one of five 41-seat examples on PSU/15 chassis—the first had been a 1952 Show exhibit. Southdown evidently favoured the previous Ambassador front-end, for the 1954 examples were of a modified design, NUF 76 being shown (below) after conversion from 41-seat in 1956 to 26-seat 'tourer' No. 1836. These were on air-pressure braked PSU1/16 chassis.

(Above) A further Beadle integral coach exercise was the 41-seat Beadle-Commer using new mechanical units, including the unusual Commer TS3 three-cylinder opposed-piston two-stroke diesel engine with its characteristic 'snarling' sound. Number 3 (RUF 103) was one of the first, dating from 1956—20 more came in 1957.

Reaction against the sharp increase in weight of most early underfloor-engined models as compared to their predecessors led to the development of lighter models, among which the Leyland Tiger Cub was to form a major part of Southdown's fleet from the mid-'fifties. It employed a horizontal 5.76-litre six-cylinder engine known as the O.350, and with the reduced weight gave better economy than the Royal Tiger, though at the cost of more noise. Southdown's first PSUC1/2 coach examples arrived in 1955, starting a renewed 1000-up fleet number series. Number 1077 (SUF 877) with Beadle body was representative of 130 placed in service up to 1958, some operating until 1970.

Despite the publicity inevitably concentrated on the hitherto unfamiliar underfloor-engined single-deckers, Southdown, like most other companies, continued to take delivery of large numbers of front-engined double-deckers. Some 112 PD2/12 models with bus bodywork were added to the fleet between 1951 and 1957, taking fleet numbers 701 upwards, following from the 1950 coach on similar chassis, but far more successful. The first 54 had Leyland bodies of the Farington type standardised for this chassis but Southdown's specification was unusual in the continued choice of half-drop windows instead of the sliding type by then more common. From 1952, platform doors were also standardised, with, at first, a 58-seat capacity—No.750 (MCD 750), dating from March 1953 and seen in Worthing on the long-established 31 service to Portsmouth, was one of the final batch. Leyland bus body building ceased the following year.

Ten PD2/12 models went to Northern Counties for bodying. Number 756 (MUF 456), new in June 1953, is seen (below) also on route 31, but bound for Brighton. Subsequent batches were bodied by Park Royal and then in 1956, Beadle bodied twelve to Park Royal design, including No. 777 (RUF 178), (below right).

(Right) The Guy Arab with 6LW Gardner engine was again in favour in 1955-56, when 48 were delivered with Park Royal bodywork, including No. 518 seen here. This time they were on Mark IV chassis, Southdown being among operators which rejected the standard 'tin front' version in favour of the traditional radiator; the projecting bonnet evident on No. 502 of 1948, visible in the background of the view above right, had gone. The Company's last 24 PD2/12 buses delivered in 1956-57 had East Lancs bodies to a similar but not identical style, as shown by No. 790 (below).

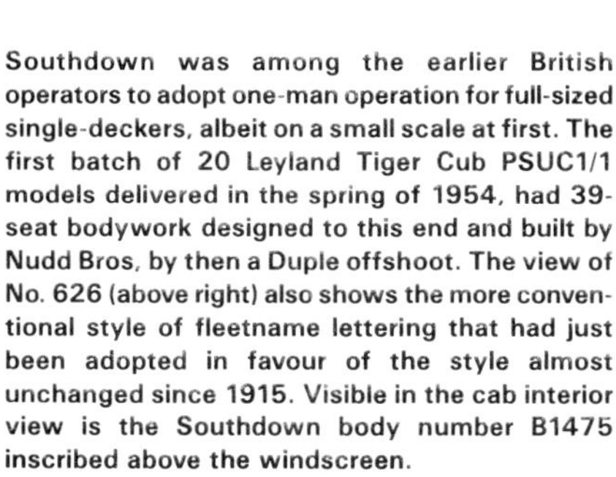

Southdown was among the earlier British operators to adopt one-man operation for full-sized single-deckers, albeit on a small scale at first. The first batch of 20 Leyland Tiger Cub PSUC1/1 models delivered in the spring of 1954, had 39-seat bodywork designed to this end and built by Nudd Bros, by then a Duple offshoot. The view of No. 626 (above right) also shows the more conventional style of fleetname lettering that had just been adopted in favour of the style almost unchanged since 1915. Visible in the cab interior view is the Southdown body number B1475 inscribed above the windscreen.

(Above) Surviving long enough to be still in the fleet when the Tiger Cubs were arriving were some of the smaller Leyland Cub buses of the late 'thirties. Number 25 (ECD 525) was one of eight KPZ2 models with Park Royal 20-seat bodywork from 1937. Seen here in 1955, it looked in good order except for a somewhat mangled lifeguard rail and was one of four to survive until 1956.

(Left) The 25 ECW-bodied Leyland Tiger PS1 coaches of 1947 were switched to bus duties in 1954-57, and renumbered in a series beginning at 675 in order of conversion. The first eight, including the former No. 1238 (GUF 738) seen in its new guise as 682, received bus seats (in this case for the original total of 31 passengers) and larger front destination blinds. Number 682 was withdrawn in 1961.

The Leyland Royal Tiger buses with East Lancashire bodywork of 1952-53 were rebuilt from rear-or centre-entrance to front-entrance layout to permit one-man operation between 1958 and 1961, No. 1502 being seen here. It was to be among the last of the type to remain in service, being withdrawn in 1968.

By then, however, the emphasis had shifted towards lighter vehicles with the Tiger Cub, first as a bus in 1954-5 but over a longer period as a coach, of which 130 examples with Beadle bodywork had entered service by 1958. Beadle had, incidentally, supplied 50 of its conversions based on Leyland TS8 running units with integral coach bodywork in 1952-4.

At the other end of the scale, the 30ft-long PD3 version of the Titan began to enter service in the fleet from 1958. All had forward-entrance Northern Counties bodywork and deliveries continued steadily until 1967, totalling some 285 including 30 convertible open-toppers. The design was virtually exclusive to Southdown, although derivations were supplied to some independent operators, and for many years this, combined with the large numbers, helped to maintain the air of 'something different' evident whenever one entered the Company's territory. It may be said that the change of livery to NBC standard from the early 'seventies did not sit too happily on this design.

Meanwhile, two of Southdown's traditional suppliers were nearing the end of their association with buses and coaches. Beadle had quite a swan song, for, in addition to the Tiger Cub coaches, and TS8-based integrals, there were a final dozen double-deckers on PD2 chassis in 1956, and 25 Beadle-Commer lightweight integral coaches in 1956-7. The Beadle-Commers were mechanically similar to 30 Avenger chassis with Harrington bodies placed in service in 1959-62, both batches having the TS3 two-stroke diesel engine made in the old Tilling-Stevens factory in Maidstone—an unusual choice for a BET company.

continued on page 88

There are many parallels between Ribble and Southdown, as fellow major members of TBAT, BET and NBC with a strong affinity for Leylands; as can be judged from a comparison between this volume and No. 2 of the same series, on the Ribble concern. Both companies turned to the Titan PD3/4 with full-fronted forward-entrance bodywork for their first 30ft.long double-deckers. This length had become permissible for double-deck buses in 1956 but the first Southdown example was No. 827 of May 1958, although numerically the last of the first batch of fifteen. Northern Counties obtained the body contract, the seating capacity being a restrained 69. Unladen weight was 8tons 3cwt 3qr, rather less than some of the Royal Tiger coaches of the early 'fifties. There was a certain irony in the use of what was clearly a modified 27ft. structure, with short bay amidships, for what was to be Southdown's largest single group of double-deckers, comprising 285 vehicles. Number 843 (below) was the first vehicle of the third batch, new in November 1959; note the various measures to increase cab ventilation and revised opening window layout generally.

Full-width front-ends had become fashionable, largely as a result of the popularity of underfloor-engined single-deckers. All of the 23 Beadle-bodied Leyland PS1 coaches of 1948-49 were rebuilt in 1954-55 with full fronts of similar design to the Beadle-Leyland integral coaches. Number 1291 is seen in Victoria Coach Station in this form. Beadle also similarly rebuilt the 40 Duple-bodied PS1 coaches that had entered service in 1948-49. Although judged desirable to give the coach fleet a modern image, it seems doubtful whether passengers benefitted from the change, as removal of the front bulkhead made it more difficult to keep the sound of the rather noisy 7.4-litre engine from the interior.

Delivery of Beadle-bodied Leyland Tiger Cub PSUC1/2 coaches continued from 1955 to 1958. Numerically the last, No. 1129 (UCD 129), dating from January of the latter year, was painted in the blue livery of the Linjebuss concern and used exclusively for that Swedish operator's tour of Britain—seating capacity was 32. It remained on this duty until January 1964 when repainted in Southdown livery and reseated to 41. It was withdrawn with the rest of the batch of ten in 1970.

Although nominally a new chassis make to Southdown, apart from acquired vehicles, the fifteen Commer coaches of 1959 were effectively a continuation of the Beadle-Commers of 1956-57, the Avenger IV chassis having the TS3 engine and other mechanical units in common. More noteworthy was a return to Burlingham as body supplier after some 23 years. The 35-seat design was one of the later and less distinguished variants of the Seagull design. Most were sold around 1971, including No. 36 (VUF 936) seen here.

A second batch of fifteen Commer Avenger IV models arrived in time for the 1960 season, the first, No. 41 (XUF 41) being seen (below left) just after delivery in December 1959. These had Harrington bodywork, again seating 35 and also not having the elegance of their bodybuilder's best designs, though better days for Harrington were yet to come, as shown on the opposite page.

A bodybuilder intermittently involved with coaches, and which had a sudden run of success in this regard was Weymann, whose Fanfare design of the late 'sixties was supplied to several BET companies, though more usually on AEC Reliance chassis. Southdown maintained its usual allegiance, however, taking delivery of fifteen on Tiger Cub PSUC1/2 chassis early in 1960. Number 1130 (XUF 130), below, carried displays for Beacon Tours duties—like the others it remained in service until 1973.

Harrington's most successful design of the underfloor-engine era was the Cavalier. Southdown took delivery of a new touring fleet in the spring and summer of 1961 consisting of 30 coaches with 28-seat bodywork of this type on the recently introduced Leyland Leopard chassis. The initial version of this model was very like the Tiger Cub with the important difference of having the O.600 engine and synchromesh gearbox, but Southdown's batch of L2T models had the added distinction of air suspension, a rare feature at that date. Another new fleet number series was begun at 1700 and No. 1701 is seen when new at Eastbourne.

Southdown took delivery of only two batches of Leyland Tiger Cub with bus bodywork, one at the beginning and the other almost at the end of deliveries of this model. This time bodywork was by Marshall, hitherto mainly an export bodybuilder but a major supplier to BET companies in the 'sixties. The bodywork was of BET design, a new departure for Southdown, and seated 45 passengers. Number 659 was one of ten similar buses all delivered in April 1962 and withdrawn almost exactly ten years later.

Further deliveries of Weymann-bodied coaches also arrived in 1962, including the last ten Tiger Cubs added to the fleet, but in October, six of the newly-introduced PSU3 36ft-long version of the Leyland Leopard were delivered with bodywork of this make. Number 1160 (160 AUF) was an Earls Court Show exhibit, its Castilian body being a development of the Fanfare, with longer window bays and a revised front-end design. It remained in the fleet until 1974 when it was sold to Hants & Dorset.

Another 'new' bodybuilder to Southdown was Plaxton, who supplied five Panorama bodies on Leyland Leopard PSU3/3R chassis in 49-seat form in January 1964, plus a similar vehicle with 35 seats for Linjebuss tours the following month. Number 1176 (176 DCD) of the former is shown below.

Deliveries of Leyland PD3 double-deckers with Northern Counties bodies to almost identical design had continued since 1958, the total of PD3/4 examples having reached exactly 100 in mid-1961. A 1961-62 batch of 40 differed in an important respect in being on PD3/5 chassis with Pneumocyclic semi-automatic rather than synchromesh gearbox—Southdown's first departure from conventional transmission since the torque converter Titans of the mid-thirties. None were delivered in 1963, but in 1964 came 50 more, including 25 of a new 'convertible' version, with detachable top cover. The basic design was little changed and the seating capacity remained unaltered, with 30 downstairs and 39 on top, but Southdown decided to revert to synchromesh gearboxes and so these were PD3/4 models once again. The open-top buses were given a new series of numbers beginning at 400, directly replacing the last survivors, themselves converted to open-top, of the wartime Guy double-deckers which had been the previous occupants of this series. Number 411 (411 DCD) is seen above bound for Arundel on the 102 coast road route on Worthing's unchanging sea front. The view (left) shows the method of storing the detachable roofs.

Further PD3/4 double-deckers with Northern Counties bodywork of basically similar design were delivered until 1967. The 1964 fixed-top buses having taken the fleet numbers to 977, a new series had been started for the 1965 batch, in almost exactly the same way as had occurred when a previous generation of Leyland Titans had reached the identical number in 1934, to avoid clashing with 1000-series Leyland coaches. However, the vehicle shown, GUF 250D, was not No. 250, the first bus of the new series, but No. 315, one of two 'special' versions of the body with experimental heating and ventilating systems (the other being No. 257 of 1965). Number 315 was a Northern Counties exhibit at the 1966 Earls Court Show, its curved glass windscreen and wide side windows not suiting this basic body design as happily as the rear-engined models on which they were usually seen.

The 36ft-long Leyland Leopard with synchromesh gearbox was Southdown's choice of single-deck bus from 1963 to 1968. A total of 135 were delivered, all having bodywork to the BET Federation's design, though the construction was split between Marshall, with 100 examples, Weymann (20) and Willowbrook (15). The Weymann batch supplied in the winter of 1965-66, including No. 158 (EUF 158D), seen above right, were among the last bodies to be built before that concern's Addlestone works was closed. The 1963 deliveries seated 51 but from 1965 the capacity was reduced to 45, this being the maximum under which one-man operation was possible under the terms of a national trade union agreement then in force. Southdown made a virtue out of necessity by installing large luggaage pens as shown above.

(Above right) Early in 1968, delivery began of 40 Bristol single-deckers, the first examples of this make ever purchased. The BET group had been impressed with the success of Bristol's RE-type rear-engined single-deckers and placed orders soon after the vehicles of this make again became available on the open market. Southdown chose the shorter-length RESL6G bus model and the batch was noteworthy for having the Gardner 6HLW engine, the hortizontal version of the familiar 6LW, rather than the more usual 6HLX, and synchromesh gearboxes, most examples by then having the semi-automatic type, though the latter was soon to become standard on the major types in the fleet. Bodywork was by Marshall, to BET design and seating 45.

Harrington was another manufacturer nearing the end of its coachbuilding days, and in October 1964 Southdown took delivery of its last batch, the 41-seat touring coaches to the final Grenadier design on Leyland Leopard PSU3/3RT chassis. Although on chassis intended for 36ft. (11-metre) bodywork, they had shorter than standard rear overhang. The first vehicle, No. 1754 (BUF 154C), is seen in the winner's parade at the 1965 British Coach Rally, having been judged runner-up to the overall concours d'elegance winner. There had been Harrington bodies from the beginning of Southdown—virtually a half-century of association.

Deliveries of Leyland Leopards after 1968 were all either coaches or, in the case of 30 PSU3/1RT models delivered towards the end of 1969, dual-purpose vehicles. These were bodied by Northern Counties to an unusual design having the front end basically to BET outline but with the main structure built to a considerably higher level, the rear end being of a style evolved by the bodybuilder. Seating capacity was 49 though two seats from that total were convertible to a luggage pen used when on bus duties. The original livery was two-tone green, the darker colour being, unusually, in the form of a continuous waistband— they were later repainted in standard green and white NBC dual-purpose style.

The formation of the National Bus Company and the automatic transfer to it of both the Southdown and Brighton Hove & District companies was marked by what amounted to 'instant' merger of the latter with the former, taking effect on 1st January 1969, the same day as NBC itself became operative. At first BH&D was operated as a separate section within Southdown, the all-Bristol fleet retaining its own red and cream livery, which had been the subject of an agreement with Brighton Corporation, which also used the same colours. This scene of June 1969 shows what at first glance seems a typical local Brighton bus scene, with BH&D and, on the extreme left, Corporaton buses in the common livery they had shared since 1939. However, the BH&D buses now carried the Southdown name as legal owners and their fleet numbers were the originals plus 2000 to avoid confusion. The almost new Bristol VR-type rear-engined double-decker (to be precise, a VRTSL6G) on the right, No. 2093 (OCD 763G), had in fact been delivered after the transfer, though ordered by BH&D. The 1965 Bristol Lodekka on the left, No. 2068 (ENJ 68C), was of type FS6B, with Bristol engine, both vehicles having ECW bodywork, as had all ex-BH&D buses taken over. (For full details of the BH&D story, see *British Bus Systems No. 4*, also published by TPC.)

In addition to Bristol VRT double-deckers added to the main Southdown fleet from 1970, the numbers of rear-engined double-deckers were expanded by the arrival of Daimler Fleetline models although as with the VRT, the first delivery was allocated to BH&D. Green and cream Daimlers were delivered from October 1970, and although the first batch of fifteen had Northern Counties bodywork, the second fifteen, based on Leyland-engined CRL6 chassis and placed in service in May 1972, had Eastern Coach Works bodies, basically similar to the contemporary VRT style but built to conventional highbridge dimensions rather than the VRT's lower overall height. They also had 'bustle'-type engine compartments, radically altering the rear-end appearance. Numbers 385 and 395 are seen in Chichester bus station when almost new.

Harrington also had a final fling, being responsible for 52 touring coaches based on its successful Cavalier design on the newly-introduced Leopard chassis in 1961-3 and a final ten on the 36ft-long version of the same model in October 1964 before following Beadle into concentration on private car activities. The first 36ft. Leopards came in 1962 and between then and 1973, some 349 were placed in service, just under half with bus bodywork to the standard BET design of the period by Marshall, Weymann and Willowbrook.

The Bristol era began at Southdown early in 1968, just after BET had agreed to sell its British bus-operating interests to the State and hence leading to Southdown's incorporation within the National Bus Company from the following January. However, the first order, for 40 RESL6G single-deckers, was one of quite a number placed by BET companies after Bristol chassis came back on to the open market, when it was widely agreed that the RE-series were the best proposition available among rear-engined single-deckers. This batch and some RELL6G delivered the following year had Marshall bodies to BET specification, though ECW-bodied Bristol VRT buses were soon to join them, as rear-engined double-deckers were introduced.

The take-over of Brighton Hove & District Omnibus Co Ltd, also in January 1969, brought a ready-made all-Bristol, all-ECW fleet under the Southdown wing, mainly composed of Lodekka and K-type double-deckers but also including ten RESL6G single-deckers—see the companion TPC volume *Brighton Hove & District* for full details. Some Daimler Fleetlines on order were allocated to BH&D, which at first remained a separate section, and, for a couple of years, the intake to Southdown's 'own' green-painted fleet was a blend of former BET and Tilling group practice. Northern Counties and, on the coach side, Plaxton and Duple bodywork was fitted mainly on Leyland chassis, with Bristol tending to keep its affinity with ECW.

The ubiquitous Leyland National single-decker came on the scene in 1973, but there was an unexpected echo of BET practice when Park Royal supplied the bodywork for Southdown's first Leyland Atlanteans in 1974—ultimately 41 arrived, together with six similar vehicles bodied by Roe. However, the usual NBC combination of Bristol VRT

For a brief period, appropriate buses carried a Southdown-BH&D fleetname in the early days of the NBC corporate livery. This photograph taken at Manor Hill in October 1973 shows a vehicle which had already had a complex history. It had originaly been delivered to Scottish Omnibuses Ltd, being one of an early batch of 25 Bristol VR double-deckers and unusual in being of the rare 33ft-long VRTLL6G type, the ECW body originally seating 83. In February-March 1973, it had been one of eight of the batch which were supplied to Southdown in exchange for the eight most modern ex-BH&D Bristol Lodekka FLF6G models dating from 1967. The seating capacity was reduced to 74 before entering service, this one becoming No. 543. The following May, the BH&D lettering was deleted from the fleetname. The vehicle, with others of the type, was sold in 1980, passing to Eastern National for driver training.

Like all other operators in the NBC group, Southdown changed over to the corporate livery style beginning in September 1972, existing vehicles gradually going into the new colours as repainting became due. The Northern Counties-bodied Leyland PD3 double-deckers were among the types that looked less at home in this style, partly because of the now redundant moulding round the lower waistline; painting the front grille surround in some cases, as on No. 266 seen here in Brighton in 1977, did not help. Even so, those reliable vehicles were still a familiar sight.

NBC decided to depart from its general policy of standardisation on Bristol-ECW double-deckers to allow expansion of the 1974 programme and orders were placed for a total of 174 Leyland Atlantean AN68/1R buses with Park Royal bodywork, all being allocated to former BET companies. Southdown received 41, all with 73-seat single-door bodywork, buses of a type which might well have been chosen if BET had continued to retain its British bus-operating subsidiaries. The Company thus received its first Atlantean models, after rather pointedly avoiding the type in the 'sixties, but the then recently introduced AN68 version was, by common consent, a much more reliable machine than the previous PDR series and indeed possibly the best of its generation. The body was of a style that had been developed for London Country Bus Services in 1972 though the basic design was Park Royal's standard of the time. The 701-up series of numbers previously occupied by PD2/12 buses was an appropriate choice for a new batch of Leyland double-deckers, No. 724 being shown here.

Early batches of Leyland Nationals—see page 52—had been of the single-doorway type, but in 1977 deliveries of 44-seat two-doorway buses were made, No. 60 being one of a batch of sixteen following another of ten supplied earlier in the year, and seen here in Hangleton. The revisions to registration index marks brought some unfamiliar newcomers, FG having hitherto been used by Fife but from 1974 reallocated to Brighton Local Vehicle Licensing Office.

Despite the general standardisation, Southdown broke new ground with the convertible open-topped version of the Bristol VR. Number 596 was one of ten VRTSL6G models with Gardner 6LXB engines and 70-seat ECW bodies, also unusual in being of two-door layout, delivered in 1977. The registration letters in this case gave a not inappropriate Brighton Hove & District flavour as the NJ mark, hitherto issued by East Sussex and hence one of those asssociated with BH&D buses, was another of those now used by Brighton LVLO. Certainly these vehicles could be counted as worthy successors to BH&D's pioneer efforts with the 'modern' seaside open-topper.

Yet another unfamiliar registration added to the 'traditional' CD and UF marks inherited by Brighton LVLO from the old borough taxation office was WV, previously used by Wiltshire and hence once associated with Southdown's former 'country cousins', Wilts & Dorset. This Bristol VRT carrying a registration with such a combination is No. 610, one of a further batch of convertible open-top buses, but this time of single-door layout and seating 74. It is seen with top in place, and distinguishable from a normal fixed-top ECW-bodied VRT only by the slightly different moulding layout at upper-deck waist and the lifting eyes on the edges of the roof. Photographed in Southampton in September 1984, it was bound for Southsea on the X16 limited stop service.

Deliveries of Leyland Leopard coaches continued through the 'seventies. Eight PSU3E/4 models purchased for the 1978 season had Duple Dominant Express 47-seat bodywork and No. 1297 is seen in its original standard National white coach livery in Pool Valley, long familiar as Southdown's bus terminal in Brighton, but now mainly a coach station, the buses having moved to Churchill Square.

By the 'eighties there was a growing reaction against the somewhat bland effect of the unrelieved white National livery and No. 1294 of the same batch as the coach shown above is in South Coast Express style, one of the variants used for limited stop services, based on a stripe pattern in the modern idiom similar to that used by other NBC fleets but given individuality by special lettering. This route, recently introduced, runs between Brighton and Portsmouth via Arundel.

A combination which would have seemed incredulous until recent years was the joint Green Line and Southdown Flightline 777 style, using a combination of green, gold and white, adopted for the London-Gatwick service. Four Leyland Leopard PSU3F/4 models with Plaxton Supreme IV Express 48-seat bodywork were purchased for this duty in 1981, including No. 1343 seen here.

Appealing to those with more reserved tastes is the once-familiar green, complete with MacKenzie-style script fleetname. Number 1835, one of the Leopard PSU3B/4R coaches with Plaxton Panorama Elite bodywork dating from 1971 restored to this style some ten years later, is seen in unfamiliar surroundings in September 1984, at Dewsbury bus station on a National Express working to Manchester. Despite the rain, the 13-year old coach looked impressively smart, only the absence of wheel trims falling short of the traditional high standards.

double-deckers, Leyland National single-deckers and Leyland Leopard coaches was soon to be dominant, accounting for over 80 per cent of the fleet by 1982. Even so, there were variations, such as the 40 Ford R1114 Duple-bodied coaches added to the fleet in the mid 'seventies, and an old tradition revived in modern guise from 1983 is the use of the Leyland Cub buses of the present Terrier-derived generation with an imaginative body design to suit able-bodied or disabled passengers built by Reeve Burgess.

Quite a long list of other makes or models not already mentioned have figured in the fleet as a result of acquisitions of other operators' business, mostly to disappear quite quickly. Among them are the following:-ADC 416; AEC B, Y and Regal; Berliet 14-seat; Caledon; Chevrolet LQ; Commer WP1; Crossley 14-seat; De Dion 14-seat; Dennis 2½, 3 and 4 ton, 30cwt, C, G, GL, E, F and Lancet; Durham-Churchill; Ensign; Fiat 14-seat; Ford T; Gilford; Graham-Dodge; GMC; Guy B; Lancia; LGOC X-type; Maudslay ML-type; McCurd; Milnes-Daimler; Morris Commercial; Napier; Romar; Scout; Straker-Squire; Studebaker; Thornycroft (various) and Unic.

A reduced need for numbers of buses as services have been re-cast to minimise operating costs, coupled with an existing fleet mostly composed of modern and standardised types, has meant that comparatively small quantities of new vehicles have been purchased in the last few years. The emphasis has been on coaches, though ten Leyland National 2 models helped to modernise the single-deck bus fleet in 1982, followed by the four special-purpose Leyland Cub buses for the County Rider services. The underfloor-engined Leyland Tiger has been the basis of two batches of 50-seat coaches, five with Plaxton Paramount bodywork arriving in 1983 and six Duple Laser versions in 1984, as well as the special Duple Caribbean-based executive coach 'The Statesman' illustrated below. Latest news as we go to press is the arrival of the first two of four more Tiger 245 models with the latest Caribbean II body, Nos. 1012-15, for the Flightline service.

It will be interesting to see how Southdown's fleet develops in response to the changing operational needs of the late 'eighties.

The Bristol VRT with ECW body steadily increased in numbers, over 200 being in service by 1982, when the model had become the most numerous single type in the fleet. Apart from the variations of body style, some examples are distinguished by various non-standard liveries. Number 274, a single-doorway 74-seat example with Gardner 6LXB engine, dating from 1981, is seen in Pool Valley wearing the rather wordy Stagecoach 700 Coastliner style, the basic message being that this is the modern equivalent of the once-famous 31 Brighton-Portsmouth service.

Following the success of the prototype County Rider shown on page 55, three more vehicles on Leyland Cub chassis were delivered in 1984. The Reeve Burgess 30-seat bodywork is similar in principle to that on the earlier bus, with the tail-lift for wheel-chair passengers at the rear, but the front-end design has been revised with a separate destination panel above the windscreen. The vehicle shown is No. 802.

Early in 1984, Southdown took delivery of a Leyland Tiger with 245 bhp engine and Duple Caribbean bodywork without seats and in plain white paintwork so that Portslade works could complete it to a top-grade executive coach specification. It duly emerged with individual armchair reclining seats for 21 passengers; a servery incorporating cooker, sink, refrigerator and bar; toilet and many other minor refinements. It was numbered 1184 and registered A184 EWV but is better known by its name 'The Statesman' carried in large letters on the brown, gold and white livery.

Southdown Surprises

As well as sizeable numbers of familiar types of vehicles, one of the Company's many appeals that account for its strong following among enthusiasts over the years has been an element of the unexpected.

The unsuspecting visitor to the more scenic parts of the country was occasionally liable to be taken unawares in the late 'thirties and again in the post-war period by the sight of a gleaming green Leyland Tigress from the six examples in Southdown's touring fleet. Among British operators, only Devon General, with ten, was a fellow user of this model, with its fire-engine overtones. This photograph, taken around 1947-49, conveys the character of these vehicles more graphically than the official portrait of 1936 on page 68, the Burlingham body having altered little in the meantime. They were all withdrawn as unacceptably outdated and sold for scrap in 1952, being replaced by the new Royal Tigers, even though it is unlikely that their condition had been allowed to deteriorate appreciably from that shown—sadly in those days, preservation of old coaches was virtually unknown.

The Company's first two Leyland Cub models of 1933 were numbered 1 and 2, and had Harrington 14-seat touring-coach bodywork on KP2-type chassis. Number 2 (ACD 102) is seen in the early 'fifties on a South Coast Express relief working—by then, seating had been increased to 20.

(Above right) rarities in several ways, Southdown's six Leyland Titan TD1 buses rebodied in 1943 were the only survivors left in the fleet from the 120 covered-top TD1 models dating from 1925-32. Park Royal built only nine of its output of over 1,000 wartime utility bodies as replacements, and these six were among them. They, together with four out of five TD2 models given Willowbrook lowbridge utility bodies, also received Gardner 5LW five-cylinder oil engines. This was not uncommon practice in some fleets, notably within the Tilling group, at the time, but unique within Southdown and it is noteworthy that they reverted to petrol around the end of the war, the seating capacity also being reduced from the usual wartime highbridge 56 to 52, in line with most of the Company's Titans at the time. Number 876 (UF 6476), dating from 1930, survived until 1952.

The idea of converting wartime utility buses to open-toppers became quite common practice in seaside fleets in the 'fifties, but Southdown was a pioneer in this field and made a particularly effective job of subtly modifying the bodywork, to give them a smart appearance. Number 466 (GUF 166) was one of a batch of Park Royal-bodied Guy Arab II buses dating from 1945 most of which became open-toppers, in this case in November 1950. Originally 5LW-engined, it received a 6LW unit in 1957 and was one of the seventeen open-top conversions of wartime Guys to survive until replaced by new Leyland PD3 buses in 1964.

APPENDICES

1

COMPANY OFFICERS

CHAIRMEN

W. F. French	1915-1925	D. S. Deacon	1971-1972
S. E. Garcke	1926-1946	F. K. Pointon	1972-1976
R. P. Beddow	1946-1968	I. Dalton	1977-1981
W. M. Dravers	1968-1968	D. L..Fytche	1981-1984
S. J. B. Skyrme	1968-1970	J. B. Hargreaves	1984-

DIRECTORS

J. J. Clark	1915-1928	S. J. B. Skyrme	1968-1970
A. D. Mackenzie	1915-1944	F. Harrison	1968-1970
A. E. Cannon	1915-1952	D. S. Deacon	1968-1973
Sir James Bradford	1915-1924	G. Duckworth	1969-1972
W. S. Wreathall	1915-1918	F. Paterson	1970-1972
W. S. Wreathall	1919-1924	I. R. Patey	1971-1972
T. Wolsey	1917-1942	F. K. Pointon	1972-1976
C. S. B. Hilton	1918-1919	L. S. Higgins	1972-1973
S. E. Garcke	1924-1948	L. S. Higgins	1975-1977
J. H. Infield	1924-1942	G. C. Smith	1972-1975
A. J. Clark	1928-1954	G. C. Smith	1977-1979
Lt-Col G. S. Szlumper	1930-1937	C. R. Buckley	1973-1975
R. G. Davidson	1930-1946	G. M. Newberry	1975-1975
J. C. Chambers	1937-1939	J. A. Birks	1975-1976
J. C. Chambers	1944-1949	J. A. Birks	1984-
C. W. G. Elliff	1939-1944	P. H. Wyke Smith	1976-1981
R. P. Beddow	1942-1968	I. Dalton	1977-1981
H. A. Short	1946-1950	A. M. Sedgley	1977-
J. W. Womar	1946-1956	D. W. Glassborow	1978-1981
H. E. Osborn	1949-1969	R. G. Roberts	1978-1979
W. H. F. Mepstead	1950-1960	F. E. Dark	1979-1984
E. Infield-Willis	1952-1972	G. R. Brook	1981-1984
Capt. The Rt. Hon		Miss K. H. Mortimer	1981-1983
Lord Teynham	1954-1969	D. L. Fytche	1981-1984
R. W. Birch	1956-1967	G. Carruthers	1981-1983
C. P. Hopkins	1960-1964	B. C. Sellars	1983-
G. L. Nicholson	1964-1968	J. B. Hargreaves	1984-
W. M. Dravers	1967-1968	J. M. Bodger	1984-

GENERAL MANAGERS

A. E. Cannon	1915-1915	G. Duckworth	1966-1972
A. D. Mackenzie	1915-1919	L. S. Higgins	1972-1972
A. E. Cannon	1919-1947	G. C. Smith	1972-1975
A. F. R. Carling	1947-1954	J. A. Birks	1975-1977
A. S. Woodgate	1954-1961	A. M. Sedgley	1977-
S. J. B. Skyrme	1961-1966		

TRAFFIC MANAGERS

A. D. Mackenzie	1915-1938	E. G. Dravers	1969-1972
A. H. Burman	1938-1946	M. Rourke	1972-1975
F. W. Sellwood	1946-1954	D. J. R. Bending	1975-1980
G. Duckworth	1954-1966	P. M. Ayers	1980-
J. C. Clymo	1966-1969		

CHIEF ENGINEERS

A. E. Cannon	1919-1926	H. R. Lane	1954-1958
R. G. Porte	1926-1933	W. G. A. Hall	1958-1968
R. E. Dunhill	1934-c1936	C. E. Clubb	1969-1973
No appointments thereafter until:		S. J. Brown	1974-

COMPANY SECRETARIES

F. Smith	1915-1924	B. C. Sellars	1967-1972
L. G. Hopkinson	1924-1948	P. J. Harmer	1972-1975
C. A. Sleap	1948-1949	D. H. Wilks	1975-
H. G. Turner	1949-1967		

AREA MANAGERS *

Portsmouth

F. Bartlett	1923-1929	F. A. Price	1956-1958
W. A. Budd	1929-1932	J. Armstrong	1958-1963
J. B. Chevallier	1932-1940	R. J. Dallimore	1963-1970
A. F. R. Carling	1940-1944	W. Wheeler	1970-1981
S. C. Hollis	1945-1947	M. W. Parkes	1981-1985
G. R. Wakeling	1947-1956		

Brighton

H. E. Hickmott	1915-1915	J. A. Birks	1964-1967
F. J. Mantell	1916-1932	G. B. Dodd	1967-1970
S. Smith	1932-1937	A. C. R. Jones	1970-1972
W. A. Stace	1938-1956	P. J. S. Shipp	1972-1976
R. J. Dallimore	1956-1963	A. Bishop	1976-1978
P. C. Hunt	1963-1964	P. S. Williams (BATS)	1978-1985

Eastbourne

S. A. Turner	c1920-1927	E. Thompson	1959-1961
S. C. Hollis	1927-1945	J. A. Birks	1961-1964
L. B. Siedle	1945-1959	M. S. Hicks	1964-1966

Combined with Brighton (1967) as Eastern Division.

Chichester

W. G. Dowle	1926-1948	Combined with Worthing as Central Division.	
E. H. Cooksey	1948-1950	J. H. Green	1970-1975
F. A. Price	1951-1956	M. W. Parkes	1975-1978
F. H. Reeves	1956-1960	A. Bishop	1978-1984 (Operating Officer for whole company except Portsmouth and Brighton Area Transport Services)
P. C. Hunt	1961-1963		
J. H. Green	1963-1970		

Bognor Regis

A. Davies	1915-1915	E. H. Cooksey	1920-1948

Combined with Chichester (1948)

Worthing

E. James	1922-1942	E. A. Butcher	1963-1967
W. J. Cooper	-1953	P. A. Townley	1968-1970
R. E. Lephard	1954-1962		

Combined with Chichester (1970) as Central Division.

* some called 'Traffic Superintendents' in early years and following re-organisations 'Divisional Managers' and 'District Managers'.

Managers of Bus Divisions from 1.3.85

Hampshire Division — Mr R. M. Higgins
West Sussex Division —
Brighton & Hove Division — Mr R. W. French
East & Mid Sussex Division — Mr I. J. McAllister

2

DEPOTS, GARAGES AND DORMY SHEDS

In many instances, particularly in the early days, stage-carriage vehicles were parked in public-house yards prior to covered accommodation being provided. In the following lists, code letters, if allotted, are given in brackets.

PORTSMOUTH [P]	1920-
Alton*	1955-1974
Boarhunt	1935 only
Clanfield	1929-1972
Emsworth	1919-1980
Fareham	1941-1975
Hambledon	1935-
Havant	1980-
Hayling**	1935-
Leigh Park†	1958-
Petersfield	1924-
Portchester	1941-1949
Purbrook	1928-1929
Warsash	1923-1924
Warsash	1929-1971
Wickham	1936-1971
BRIGHTON [A]	1915-
Bolney	1918-1965
Burgess Hill	1922 only
Chelwood Gate	1922-1967
Cowfold	1923 only
Crawley	1959-1971
East Hoathly	1923-1966
East Grinstead	1933-1972
Hassocks	1952-
Haywards Heath	1925-
Henfield	1923-
Moulsecoomb	1957-
Lewes	1929-
Scaynes Hill	1920-1935
Wivelsfield	1931-1950
EASTBOURNE [E]	1920-
Alfriston	1929-1954
Crowborough	1949-1980
Hailsham	1957-
Heathfield	1923, 1927-1971
Seaford	1915-
Uckfield	1920-
Upper Dicker	1929-1967
BOGNOR [B]	1915-1982
Chichester (C)	1923-
Compton	1930-1964
Eastergate	1944-1961
Midhurst	1924-
Petworth	1921-1971
Selsey	1921-1964
Singleton	1924, c1927-1954
Summersdale	1924 only
Walberton	1922 only
West Dean	1924-c1927
Wittering	1923-1965
WORTHING [W]	1915-
Barns Green	1935-1954
Dial Post	1928-1966
Handcross	1920-1965
Horsham (WH)	1919-
Littlehampton (WL)	1921-1971
Pulborough	1927-1971
Steyning	1919-
Storrington	1915-

* Aldershot & District Traction Co Ltd garage.
** Ministry of Aircraft Production establishment 1939-1946
†City of Portsmouth Passenger Transport Department garage

Compiled by Southdown Enthusiast Club.

3

EXCURSION AND TOURS OPERATORS' LICENCES ACQUIRED

Arthus Davies, Bognor	1915
Thomas Tilling Ltd, Brighton	1917
W. C. Taylor 'White Heather Motor Services', Brighton	1921
E. Higgins 'De Luxe', Brighton	1923
C. G. Shore 'Royal Blue Services', Bognor Regis*	1923
South Coast Tourist Co Ltd, Littlehampton*	1924
Cavendish Coach & Car Company, Eastbourne	1925
Southsea Tourist Company, Portsmouth*	1925
Miskin & Strong, Littlehampton	1925
S. Foard, Eastbourne	1926
J. Williams, 'The Pullman Service', Hove	1927
F. Bevan, 'Golden Butterfly Motor Coaches', Brighton	1927
Noble, 'Lucille Motor Coaches', Brighton	1927
F. Rhodes, 'Bonny Doone', Brighton	1928
Unic Chars-a-Bancs, Brighton	1929
F. & J. Poole, 'Royal Red Coaches', Hove	1930
A. Potts, 'Potts Coaches', Brighton	1930
Chapman & Sons (Eastbourne) Ltd †	1932
G. H. Meaby, Brighton	1932
Pratt & Pearce Ltd 'Little Vic Motor Coaches', Eastbourne	1933
P. C. Belier, Hayling Island*	1933
W. G. Waugh, 'Regal Coaches', Brighton	1934
A. D. Jessett's 'Jessett's Coaches', Haywards Heath	1934
B. R. Roberts, Hove	1934
W. F. Alexander, 'Comfy Coaches', Horsham	1935
Rowland & Orr, Worthing	1936
C. F. Gates, 'Silver Queen', Worthing	1936
C. C. Lansdall, 'King of the Road Coaches', Worthing	1936
H. Miller, 'Miller's Coaches', Portslade	1936
F. & H. L. Pownall, Brighton	1937
S. W. Stephens, 'Sydney Motor Coaches', Eastbourne	1938
H. J. Sargent, 'East Grinstead Services', East Grinstead	1938
Hartington Motors Ltd, Eastbourne	1938
Harvey, Eastbourne	1939
French & Son, Seaford	1945
George Ewen, 'Pioneer Coaches', Petersfield	1952
C. F. Wood & Sons, Steyning	1955
Triumph Coaches, Portsmouth†	1957
Brighton Motor Conveyances (Shamrock & Rambler), Brighton	1963
A. W. Buckingham, 'Bucks Luxury Coaches', Worthing	1963

* together with stage-carriage work
† together with express-service work

4

LIVERY AND INDICIA

Until World War II, Southdown buses — officially mid-green and cream — were painted apple green and primrose with BAT dark-green wings and horizontal beading. The coaches featured BAT dark-green window surrounds and roof as well. The lower panels of stage carriage vehicles were lined out in gold with an ornate corner-piece. This was not the usual BAT 'overlapping fern shoots', but a design which appeared to owe much to Japanese family crest-motifs of the Edo period. It resembled two embryonic fish trying to shelter under the same umbrella. Just who seems to have designed that seems to have been forgotten.

Extremely vulnerable to the attentions of the cross-Channel Luftwaffe, most stage carriage vehicles and some coaches were painted grey above the waistband. Latterly, some double-decks acquired an apple green band below the upper-deck windows, whilst others received BAT dark green roofs.

After the war there was a return to the basic colour schemes, but what lining-out there was had lost its 'fish and umbrella'. From 1951, coaches were overall apple green, but during the 1960 season only, touring coaches were primrose above the waistrail.

Until the introduction of NBC's bland leaf-green and white livery in 1972, double-deckers and most saloons carried an ornate block-capital fleetname **SOUTHDOWN** in gold lined in black to the left and below each letter to give a three-dimensional effect. Mackenzie remembered his 'Swiftsure-style' cursive script in 1922 and thereafter coaches, some post-war Guys, particularly the open-toppers, and some later saloons bore their fleet-name in the same flowing style. Two-and-a-half inch fleetnumbers in black-edged gold were applied toward the rear side panelling, off and near, until the NBC era when they were replaced by standard pale grey numbers fore and aft. Fleetnames then changed to the stick-on 'most of the time' white block capital type together with 'double N' logo, the latter later changing to red and blue on a white ground which, in turn was replaced by the 'Chairman's award' motif in 1981. From 1972, most of the coaches went into the white NATIONAL livery.

Acknowledgements

Twenty years ago, in celebration of its 50th anniversary, Southdown Motor Services Ltd published its own book, **The Southdown Story 1915-1965,** and it is now something of a collector's item. This was the first publication devoted solely to the subject since the appearance of Ian Allan's 41-page booklet **The ABC of Southdown Buses & Coaches** in 1950. Curiously, bearing in mind the distinguished record of the company and its particularly large following, this book is next in line, and the first to appear case-bound in the shops. For vehicle-detail there are of course the joint PSV Circle/Omnibus Society/Southdown Enthusiast Club booklets 1PK I (1972) and 2PK II (1974), whilst passages and a chapter or two have featured Southdown in Henfry Smail's **Worthing Pageant: Coaching Times and After** (1948), several issues of **Buses Annual** and Colin Morris' **Regional History of British Bus Services, Vol. 1 — South East England**, TPC (1980).

In most other instances one has to turn to articles in newspapers and journals and for this volume the following were consulted: **Bognor Post, Brighton & Hove Herald, Bus/Bus News/Southdown Bus News, Buses Illustrated/Buses, Chichester & Southdown Observer, Commercial Motor, Daily Telegraph, Eastbourne Chronicle, Eastbourne Gazette, Evening Argus,** Brighton, **Evening Standard, Financial News, Hampshire Telegraph & Post, Hants & Sussex County Press, Hants & Sussex News, The Herald, Lancing & Shoreham Times, Leyland Journal, Licensed Victuallers Gazette & Hotel Courier, Littlehampton Gazette, Mid-Sussex Times, Modern Transport, Motor Transport, News Chronicle, Old Motor, Portsmouth Evening News/The News,** Portsmouth, **The Star, South of England Advertiser, Southern Daily Echo/Evening Echo,** Southampton, **Southern Weekly News, Stock Exchange Gazette, Sussex County Herald, Sussex Daily News, Sussex Express & County Herald, The Times, Tweenus** and **Southdown Chronicle** (1936-1972), **West Sussex County Times, West Sussex Gazette** and the **Worthing Herald.**

Much of the field work to cover the history of the company prior to the NBC era was undertaken over a decade ago with help and facilities provided by the then publicity officer, Joe Carr, and chief general manager, Geoffrey Smith — and his secretary, Miss Puttock, together with John Roberts, pushed me in the right direction. Keeping the information flowing from Southdown House for what turned out to be a tight writing-schedule was Assistant Traffic Manager, Roger French (no relative he assures me of Walter Flexman French). My frequent requests for documentation were speedily dealt with also by his secretary, Angela Ketley. My thanks to them both, for they made possible those sections which deal with the last ten years.

I am particularly grateful to Mr R. J. Dallimore — an ex-area manager of Brighton and then Portsmouth — who took the trouble to write several letters containing useful material and who wishes to set on record the part which the committee of the Southdown Portsmouth Area Sports & Social Club and its members played in the setting up and launching of the company house-magazine **Tweenus**, in whose pages over the years was safely recorded as it happened much of the commercial, social and sporting history of the company.

Among others who offered help and advice were Mr A. F. R. (Michael) Carling; Mrs Gerald Duckworth; Mr P. M. D. Stobart, traffic manager of the China Motor Bus Co Ltd, Hong Kong; Mr M. A. Farncombe, London Country's St. Albans traffic superintendent and his father, Mr Thomas Farncombe, who drove the first South Coast Express coach from Margate to Bournemouth; Mr Frank Jones of Ardingly and Mr T. Gilks of Brighton. I have also had my attention drawn to some fascinating facts by that respected writer, George Behrend — who would undoubtedly claim that I hail from the 'wrong part' of Hampshire.

Finally, I have to thank Alan Townsin for his inimitable captions to the illustrations. I'm sure Mackenzie would have liked the mechanical detail, to say nothing of the numbers, and who else could romanticise so effectively over the sophistication and curvaceous beauty of the back of a bus or — in less gallant mood — draw attention to her sagging waistline and poorly-fastened bonnet? 'Pass down the car, please — anyone else want to pay twice?'

Colin Morris
Department of Art & Design
Faculty of Education & Community Studies
Liverpool Polytechnic

A Final Word from a Passenger:-

"I want to thank everyone here at Southdown very much for the service I have had. I couldn't have managed without the drivers. Thanks to them I haven't missed a day in 25 years".

These were the words of a disabled passenger when he said farewell to his old friends at Southdown after 25 years of travelling by bus.

Bob Craiger, 43, who is spastic, has made the journey from his home in Yapton to either Brighton or Worthing every weekday for 25 years.

Before he set out on his final Southdown trip, Bob said a sad farewell to the staff who have become his friends over the years.

Before he set off for home the staff at Southdown had their own special surprise for him. The Company presented him with a silver tankard inscribed with his name, and the staff presented him with a carriage clock. Bob was also given a Southdown calendar, tie, scarf and tea towel.

Then, instead of catching his usual bus to Yapton, Bob was taken home by Southdown coach.

With acknowledgement to the Evening Argus.

PHOTO CREDITS

Southdown's own archives provided by far the largest number of photographs reproduced in this book, and the Company's assistance in making them available is gratefully acknowledged. Thanks are also due to the following contributors.

Contributor	Pages
G. H. F. Atkins	37
A. H. Barkway	23(lower),40(upper),57(lower)
A. A. F. Bell	40(lower)
D. Clark	72(upper right)
A. B. Cross	43(lower)
John Cull collection	74(upper right and left),75(upper left, centre left)
M. J. Dryhurst	85(upper centre)
J. C. Gillham	51,88(lower)
J. A. Goddard	48(upper right)
R. N. Hannay collection	35(lower),38(centre),48(lower),72(upper left, lower),74(centre),75(upper right, lower) 76(centre right, lower),77(lower),78(upper right),89(upper centre),81(all)
M. Keeley collection	53(upper),54,89(centre),90(lower),91(lower)
C. F. Klapper	25(upper), 72(centre left)
A. Lambert/A. Lambert collection	8,11,12(lower),13,16(upper right),17(centre lower),18,19(lower),21(upper),23(upper), 24(both),25(lower),26(both),30(upper),32, 46,48(upper left),57(upper left and centre), 58(upper),59(centre),60(upper),76(centre right and lower)
H. Luff	44(upper)
R. F. Mack	76(centre left)
R. A. Mills	77(upper left)
Martin Montano collection	87(upper right, lower),89(lower)
Colin Morris collection	30(lower),91(upper)
Eric Nixon	Front Cover
S. L. Poole	67(upper right)
Portsmouth & Sunderland Newspapers Ltd	27
P. Poulter	82(lower)
John Roberts collection	52(lower),88(upper),89(upper),90(upper, centre)
J. V. Short	85(upper)
H. J. Snook	17(centre)
Southdown Enthusiasts Club	87(upper centre)
Surfleet Transport Photographs	16(upper right),59,72(centre right),77(upper centre right),78(upper left),93(upper and lower centre, lower)
A. A. Townsin/A. A. Townsin collection	10,38(upper),67(centre right),73(all),77(upper right, lower centre),87(upper left, lower centre)
TPC library courtesy:	
Brush	33,64(centre)
Burlingham	84(upper centre)
Harrington	71(lower)
Hestair Dennis	78(centre left)
Leyland Vehicles/British Commercial Vehicle Museum Leyland	28,29,34(upper),36,42(upper),43(upper), 64(centre),69(centre),70(lower centre, lower 74(lower)
Albert Meek collection	5,71(centre),72(centre right),77(upper centre right),83(lower),93(upper and lower centre, lower)
Northern Counties	78(centre right)
Short Bros.	63(lower),64(upper),66(upper and centre)